Electromagnetic Levitation and Suspension Techniques

B V Jayawant

Professor of Electrical Engineering, University of Sussex

Edward Arnold

First published 1981 by
Edward Arnold (Publishers) Ltd
41 Bedford Square, London WC1B 3DQ

British Library Cataloguing in Publication Data

Jayawant, Bhalchandra Vinayak
Electromagnetic levitation and suspension techniques.
1. Magnetic suspension
I. Title
621.319'21 TK454.4.M3

ISBN 0-7131-3428-3

Printed in Great Britain by
Thomson Litho Ltd, East Kilbride, Scotland

Preface

There has been a great deal of excitement in the last decade in the field of electromagnetic suspension and levitation due to the possibilities of application to advanced ground transportation vehicles. Although in 1935 Kemper suspended a mass of 210 kg, applications only became viable with the recent rapid advances in solid state electronics. There has been an accompanying surge of published work relating to advanced ground transportation schemes using electromagnetic levitation techniques.

This book is an attempt to survey the current scene in a broad manner, covering the existing state of the available technologies and their potential, as well as some of the unsolved problems. More than one technology is now in a state where systems of passenger carrying vehicles using it appear to be perfectly feasible. The present economic climate in the western world, however, makes it virtually impossible to predict when, if at all, such schemes might see the light of day as public fare-paying systems. In any other field except that of transport, the merits of the technology, operational costs, and shortage of fossil fuels would ensure a serious attempt to harness the fruits of past ten years' labours. Instead, at the time of writing this book anyway, we appear to be proceeding unconcerned, putting more cars on the road and spending money on motorways, which will leave their mark for generations to come, when railways already exist.

In the book I have tended to avoid 'transportation' as such and to be very cautious in extolling the virtues of electromagnetically suspended or levitated advanced ground transportation schemes. I hope that this book will provide a fairly objective overview of the possibilities to any interested party. This is certainly my hope for the non-transport applications, some of which have already reached a successful state: for example, frictionless bearings and non-contact suspension in the textile area. There is a vast untapped potential for 'industrial' applications. The emphasis is therefore on references to published work, and there are many photographs of some of the exciting developments of the last decade achieved by researchers in this field. I hope to find school teachers or even school children as well as future research workers in the field of electromagnetic suspension browsing through the text.

Working in this field I have made many friends in all parts of the world and I am grateful to all those who have made material available for the book. Finally a special word of thanks to all my research staff, past and present (including research fellows, research students and technicians) and colleagues in the School of Engineering for their invaluable contribution to our research effort for the past decade.

1981 B. V. Jayawant

Contents

1

Survey of suspension and levitation techniques

1.1 Introduction

The phenomenon of levitation has fascinated philosphers through the ages and in recent times it has attracted much attention from scientists as a means of eliminating friction or physical contact. Although the area of frictionless bearings is important, it is the application of contactless suspension and levitation to high-speed ground transportation which has received the most attention, especially in the popular media. Regardless of the method employed the vehicles are described as 'Hover Trains' and any electromagnetic method is ascribed the title 'Maglev'. To indicate the author's dislike of this term, it will not be used in later chapters of this book.

Besides the aircushion principle of supporting rotating shafts or vehicles, as used in the Aerotrain in France or the Tracked Hovercraft in the UK, there are nine other electromagnetic methods of supporting moving or rotating masses[1]. These methods are as follows.

1) Levitation using forces of repulsion between permanent magnets.
2) Levitation using forces of repulsion between diamagnetic materials.
3) Levitation using superconducting magnets.
4) Levitation by forces of repulsion due to eddy currents induced in a conducting surface or body.
5) Levitation using the force which acts on a current-carrying linear conductor in a magnetic field.
6) Suspension using a tuned LCR circuit and the electrostatic force of attraction between two plates.
7) Suspension using a tuned LCR circuit and the magnetic force of attraction between an electromagnet and a ferromagnetic body.
8) Suspension using controlled DC electromagnets and the force of attraction between magnetised bodies.
9) Mixed μ system of levitation, where μ is the permeability of the material.

Some of the possible methods of suspension or levitation in this list are of mainly academic interest but others, three in particular, were the subject of intensive experimental activity in the early 1970s. It is first of all necessary to draw a distinction between those methods which use forces of attraction and those which use forces of repulsion. The former may be called **suspension** techniques and the latter **levitation**; each one of the nine proposed methods is virtually a technology in its own right. It is therefore apparent how sweeping and ill-conceived is the term Maglev.

1.2 Principles and limitations of electromagnetic techniques of suspension and levitation

Every one of the methods listed in Section 1.1 has at one time or another been the subject of enthusiastic investigation. The difficulties of achieving stable suspension or levitation are, however, highlighted by an examination of the nature of the forces in the case when an inverse square law relates force and distance. Earnshaw[2] in his paper, which is now considered a classic by all workers in the field of electromagnetic suspension, shows mathematically that it is impossible for a pole placed in a static field of force to have a position of stable equilibrium when an inverse square law operates; this fundamental calculation is known as Earnshaw's theorem. Braunbeck[3] carried out a similar analysis specifically for unvarying magnetic and electric fields, and deduced that suspension or levitation is not possible in such fields when all materials present have $\mu_r > 1$ or $\epsilon_r > 1$, but that it is possible when materials of $\mu_r < 1$ or $\epsilon_r < 1$ are introduced; μ_r is the relative permeability and ϵ_r is the relative permittivity of the material.

Suspension has been achieved using diamagnetic materials, and as a result of the effect of currents in the suspended objects. Several variants of the latter method in particular are at present the subject of extensive investigations.

1.2.1 Suspension or levitation using permanent magnets

It follows from Earnshaw's theorem and Braunbeck's analysis that stable suspension or levitation is impossible with a system of permanent magnets (or fixed current electromagnets) unless part of the system contains either diamagnetic material ($\mu_r < 1$) or a superconductor ($\mu_r = 0$), and that it is altogether impossible to achieve levitation in electrostatic fields since there are no known materials with $\epsilon_r < 1$.

Work on the use of permanent magnets began around 1890 and was at that time concerned with relief of part or the whole of the load or a rotor or shaft magnetically. In spite of the large number of patents none of this work appears to have been commercially successful. Interest in the topic seems to have languished until the 1930s and the advent of improved permanent magnet materials. The most common application of magnets of fixed strength is in the suspension of shafts or spindles; suspension of the spindles of watthour motors in order to relieve the load on the pivot bearing has been a constant target.[4–6] Other possible applications in the aerospace field are discussed in a report from General Electric (US).[7]

Recent developments in fabricating permanent magnets from high coercivity ferrite materials have once again raised interest in the idea of using such magnets for levitation of vehicles. Polgreen in 1959 was the first to propose an application in the form of a vehicle,[8–11] using newly developed barium ferrite magnets[12,13] in the repulsion mode. His was a model system consisting of barium ferrite magnet blocks fixed to the underside of a vehicle, with nylon rollers for guidance (see Fig. 1.1).

Similar proposals, including one for high speed travel in evacuated tubes across the USA, were made by others at about the same time.[14–16] In all these proposals it was assumed that barium ferrite would be cheap to manufacture in the large quantities which would be required when laying down the track. One of the advantages of using ferrites for the track is that there are no induced eddy currents whose

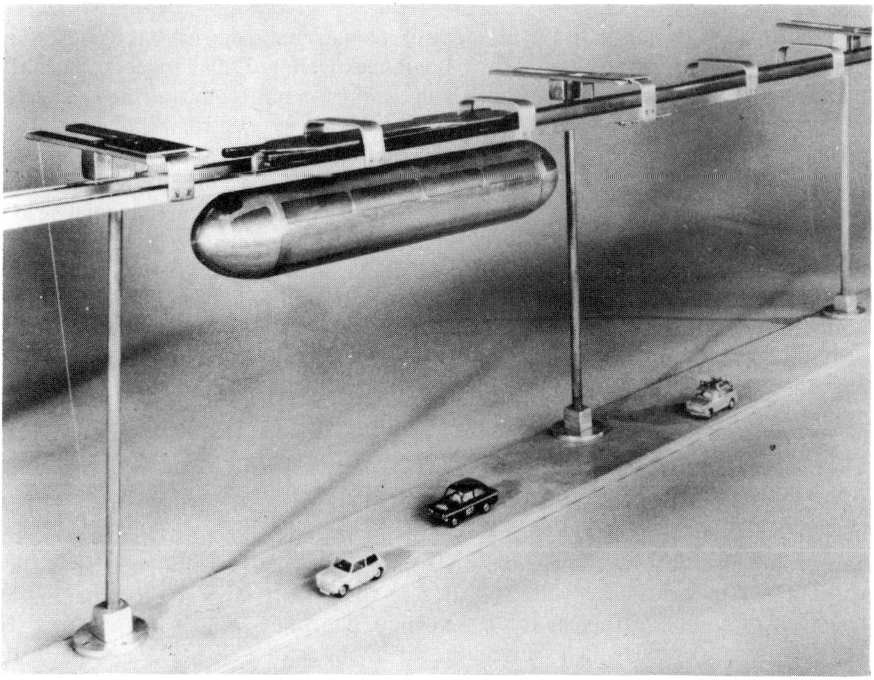

Fig. 1.1 Model of Magnerail over road centre with motor cars to scale (photograph courtesy of Mr G. R. Polgreen).

reaction would lead to either a drag force or a loss of lift. McCaig[17] and Bahmanyar and Ellison[18] have made a study of the lifting forces, configurations and track designs. However, any practical system built around these ideas would require damping in the vertical direction and guidance as well as damping in the lateral direction. This has been considered only by Voigt,[19] who proposed driving a current proportional to vertical velocity in a coil around the permanent magnets. Recently a new magnetic material has appeared, namely samarium-cobalt, which has even greater coercivity than barium ferrite. The intrinsic coercive force of these cobalt rare earth materials may be 20 to 50 times that of conventional permanent magnets, producing lifting capabilities in the repulsion mode which are 5 to 10 times greater.[20] In spite of these significant improvements in the materials available there remain many practical difficulties in the implementation of transportation schemes using permanent magnet tracks. However, for instrument applications there is a real possibility of using controlled permanent magnet schemes whereby the power consumption in the steady state can be made virtually zero.

1.2.2 Levitation using diamagnetic materials

As indicated in Section 1.1, levitation can be achieved in static magnetic fields by employing diamagnetic materials. However, even the two materials which exhibit the most pronounced diamagnetic properties, i.e. bismuth and graphite, are so weakly diamagnetic that only small pieces can be levitated.

The topic of levitation in the presence of diamagnetic materials has been studied by Braunbeck[3,21,22] andBoerdijk.[23,24] Braunbeck levitated small pieces of bismuth, measuring 0.75 x 2 mm and weighing 8 mg, and of graphite, measuring 2 x 12 mm weighing 79 mg, between specially formed poles of an electromagnet capable of producing a field of flux density 2.3 tesla. Boerdijk repeated this experiment on a smaller scale using permanent magnets and also performed an alternative experiment which involved levitating a magnetised disc of diameter 1 mm between a magnet attracting it upwards and a piece of diamagnetic material below it. His analysis concludes that it should be possible to levitate a magnetised particle of micrometric size a fraction of a millimetre above a piece of bismuth or graphite, without the aid of a surmounted magnet. A further examination of diamagnetic levitation is contained in a report by General Electric (US)[7] but the general conclusion to be drawn from the results of that work is inevitably that the phenomenon of diamagnetic levitation is of academic interest only.

1.2.3 Levitation using superconductors

Certain metals and alloys when cooled to a temperature approaching 0 K (−273°C) become superconductors. The superconducting state is indicated by the complete absence of electrical resistance, and once initiated a current will continue to flow without the presence of a voltage source in the circuit. This is accompanied by rejection of magnetic flux in the superconducting body, which is known as the Meissner effect[25] and which causes the superconductor to behave as a perfectly diamagnetic material ($\mu_r = 0$). Stable suspensions using permanent magnets are therefore possible.

The first recorded demonstration of this principle was by Arkadiev[26] in 1945, and involved the levitation of a 15 mm bar magnet over a superconducting lead plate. In the subsequent attempt to develop a cryogenic magnetic suspension gyroscope, superconducting spheres were levitated over various arrangements of electromagnets.[27,28]

A scheme for levitating a vehicle over two parallel superconducting rails was proposed by Powell[29] in 1963. Later, a second system[30] (see Fig. 1.2) was designed in which superconducting magnets were attached to the vehicle; these magnets would ride over rails without touching them, thus eliminating the need for superconducting rails. There were also proposals to support a rocket launching sledge[31] capable of speeds of 5 km s^{-1} (see Fig. 1.3) and further studies of baseline specifications[32] for passenger carrying vehicles.

An electrodynamically levitated vehicle is lifted and guided by repulsion forces between superconducting magnets on the vehicle and secondary circuits on the track (eddy currents if the track circuits are passive). The levitation is self-stabilising and clearance between magnets and secondary circuits can be larger than 10 cm. However, the stiffness and damping of the suspension are low, and the vehicle must be in motion in order to generate lift. There is therefore a minimum velocity which must be exceeded before the vehicle becomes levitated and the system is generally considered applicable only to high speed transport systems (in excess of 400 km h^{-1}). Some of the problems and drawbacks which still remain unsolved are that in addition to the aerodynamic drag on such vehicles there is an eddy current drag which is large at low speeds. This places quite a substantial burden on the propulsion systems during acceleration. The drag reduces at high speeds but in order

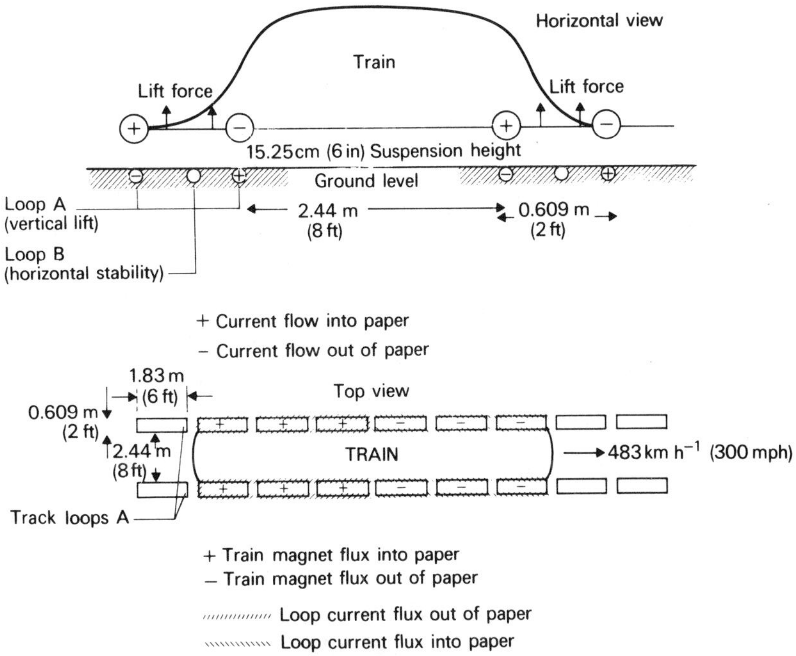

Fig. 1.2 Conceptual view of a train levitated over two parallel superconducting rails, as proposed by Powell and Danby.

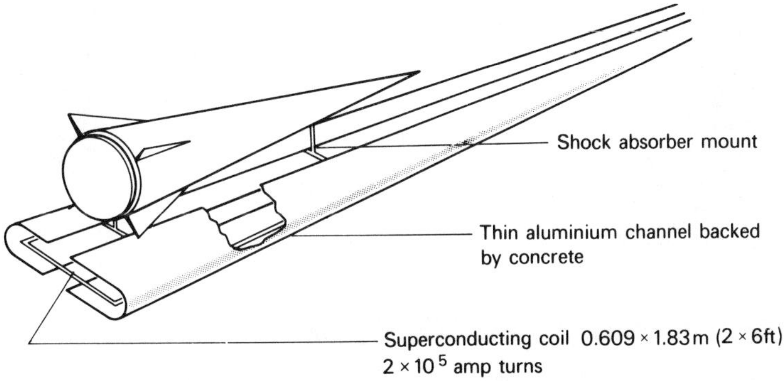

Fig. 1.3 Artist's conception of a rocket sled with superconducting magnet and track.

to get a high lift-to-drag ratio, which is a figure of merit for these systems, a large quantity of conducting material (aluminium) is required in the secondary circuit (track). At high speeds the low inherent damping coefficient of the system reduces further and can in fact become negative, which presents quite serious problems of vehicle stability in general. It has been reported that passive damping may be inadequate.[32–6] Proposals considered have been to vary the coil currents,[37] or to vary the drive to the linear synchronous motor in accordance with vertical acceleration signals fed back,[38] or in order to produce more damping at the expense of the figure of merit.

Fig. 1.4 Japanese National Railways superconducting levitated test vehicle ML-100 (photograph courtesy of Japanese National Railways).

Research in the superconducting levitation systems is quite active in Canada, Japan and England. The Japanese National Railways produced in 1972 a $3\frac{1}{2}$ tonne vehicle (see Fig. 1.4) which had a lift of 6 cm but in which guidance was provided by wheels on the sides of the guideway. A second and more advanced vehicle operating on a 20 km track was reported in 1979 to have achieved speeds in excess of 250 km.[39,40]

There were two projects of this type in the US. One was a collaborative effort between various universities and industrial laboratories under the direction of the Department of Transportation. The other, called the Magneplane project, was partly under the direction of the National Science Foundation. Both studies were largely theoretical, the experimental side involving permanent magnets.[33,37,41,42] Research in the US appears to have been suspended indefinitely since about 1975.

The research in Canada in superconducting levitation systems for high speed ground transportation using synchronous linear motor propulsion is being carried out by an interdisciplinary team of scientists and engineers from the Universities of Toronto, Queens and McGill.[43] A wheel of diameter 7.6 mm rotating about a vertical axis with a maximum peripheral speed of 100 km h^{-1} is being used to carry out full-scale tests of the propulsion, levitation and guidance systems[44] (see Fig. 1.5).

Fig. 1.5 The Canadian Universities (CIGGT) test wheel with linear synchronous motor windings in the rim (photograph courtesy of the Canadian Institute of Guided Ground Transport).

In the UK work has been in progress for a number of years[45–7] at the University of Warwick; and a 600m track has been constructed to test a small vehicle which, initially, is to be towed by a rope at speeds up to 35 m s^{-1}. This small vehicle is 3 m long and weighs 150 kg.

Studies are also being carried out by a consortium of Siemens, AEG and Brown Boveri[48,49] at Erlangen in West Germany and a vehicle has undergone preliminary tests on a 280 m diameter circular track (see Fig. 1.6). It is believed that because of the problems of guidance and eddy current drag this work has been delayed quite considerably.

1.2.4 Levitation using induced eddy currents

A force of repulsion is generated between a coil carrying alternating current and an electrically conducting surface in the proximity of the coil, so that the alternating

Fig. 1.6 Test track at Erlangen, W. Germany, for superconducting magnetically levitated vehicles (photograph courtesy of Siemens Ltd).

magnetic field of the coil induces eddy currents in the conductor. This effect can be utilised for levitation of conducting objects and one of the early patents purporting to do so is that of Anschütz-Kaempfe[50] in gyroscopic applications. This technique has also been used for simultaneous levitation and melting of specimens[51] at 10 kHz for zone refining of metals (see Fig. 1.7). This has become a standard technique for preparation of small quantities of alloys without contamination from crucibles.

Bedford, Peer and Tonks[52] describe a plate levitator in which two concentric

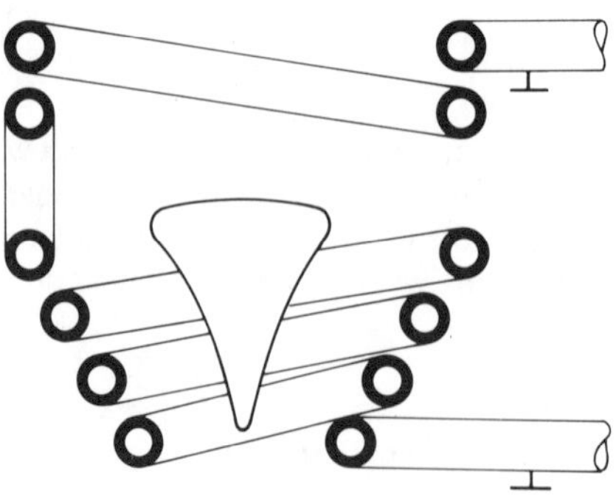

Fig. 1.7 Typical construction of a radio frequency coil system for the levitation of molten metals. The coils are water cooled copper tube. The upper (stabilising) coil is in series opposition to the lower (levitating) coil. The molten metal assumes approximately the shape and position shown.

coils carry 50 Hz currents in opposite directions and can levitate a circular conducting plate in stable conditions; several other experimental systems for levitation of plates, spheres etc. are described by Laithwaite.[53] More recently, however, due to developments in linear induction motors, particularly of the transverse flux type,[54,55] it has been claimed that such machines might be used for combined levitation and propulsion of high speed vehicles.[56] On the basis of a great deal of experimental work on relatively small models such as those in Fig. 1.8 it is suggested that, due to scaling laws for electromagnetic machines,[57] combined levitation and propulsion schemes employing linear motors for vehicles weighing in excess of

Fig. 1.8 Prototype of the model vehicle illustrating the principles of the magnetic river scheme (photograph courtesy of Professor Laithwaite).

50 tonne may have performance comparable to that of the superconducting magnet schemes. One of the advantages claimed for such schemes, termed 'magnetic rivers', is that they offer the possibility of lift and guidance where the motor necessary for propulsion is the source of such facilities. It is also claimed that for a particular thrust the secondary power input in a levitating linear motor will be the same as in a machine designed for thrust only. Obviously a great deal of work, particularly theoretical, needs to be done but it is difficult to envisage that there may be dramatic improvements to primary (reactive) power input for the large airgap operation claimed as one of its operating features, and also that the stiffness and magnitude of guidance and lift forces are adequate in the geometries examined so far for the proposed advanced ground transportation applications. Recently Eastham[58] has shown that the theoretical minimum reactive input to a 50 tonne vehicle energised from a 50 Hz supply is 30.8 MVAR for levitation at a gap of 0.1 m, compared to 3.25 MW required to propel the vehicle at 120 m s^{-1} to overcome aerodynamic drag. Eastham also shows that such a machine will have an inherently poor power factor and if on-board power factor correction equipment is used to bring it up to unity power factor there is a substantial weight penalty. The overall

efficiency × power factor product is then 0.38 for such a vehicle. However, the use of transverse flux linear induction motors (TFMs) as propulsion units only remains very promising.

1.2.5 Levitation using forces acting on current carrying conductors situated in magnetic fields

The force acting on a conductor of length l carrying a current I and situated in a transverse magnetic field of flux density B is given by BIl and the force acts in a direction normal to both the conductor and the field. Pfann and Hagelbarger[59] report having supported the molten portions of a metal rod undergoing zone melting by locating the molten portion in a transverse magnetic field and passing a current through the rod. Although the current is adjusted to give an upward force approximately equal to the weight of the molten metal, surface tension also contributes to keeping the molten zone in place. The heating of the molten zone is carried out either by induction heating or by a torch flame. Thus, unlike the eddy current levitation technique, the functions of melting and levitation are kept separate. Rods of iron, nickel and tin have been levitated by this method.

A variant of the same technique was proposed by Powell[29] for levitation of a vehicle over two parallel superconducting rails carrying a persistent current. Attached to the vehicle are two superconducting inverted troughs which ride over the rails without touching them. Levitation is effected by persistent currents flowing in the longitudinal wires of which the troughs are constructed. The troughs are designed to give the vehicle stable equilibrium both vertically and laterally. In a paper which contains technical and economic calculations and a report of preliminary experiments, Powell estimates that with a current of 300 000 A and a trough radius of 45 cm (18 inches), a weight of 4500 kg m^{-1} (34 lb ft^{-1}) could be supported. The idea does not seem to have been developed further since its publication, and a recent discouraging report about the prospects for superconducting cables[60] would suggest that it is not likely to, on both technical and economic grounds.

1.2.6 Suspension using a tuned LCR circuit and electrostatic force of attraction

An electrically conducting shaft or rotor may be held in suspension by electrostatic forces between a pair of electrodes where one of the electrodes is the body to be suspended. The suspended body and the fixed electrode form the capacitive element of a tuned LCR circuit in such a manner that the potential difference between the two electrodes increases as the distance between them increases, and vice versa, i.e. the circuit is tuned to resonance at values of capacitance less than that at the suspension gap. The electrodes must be maintained at a potential difference of several kilovolts. Applications of this principle have been investigated[61,62] for vacuum gyros. This technique does not seem to have been as extensively investigated as the one using the magnetic force of attraction in tuned LCR circuits. It is, however, almost certain that besides the problem of the very high voltages which are required to achieve suspension, this method also suffers from inherent instability due to the use of tuned LCR circuits, and the problems of providing damping and high reactive power are just as adverse as in the LCR systems employing a variation of inductance with gap width.

1.2.7 Suspension using a tuned LCR circuit and electromagnetic force of attraction

As indicated in Section 1.2.6 this method has been the subject of extensive investigation, particularly by groups at the MIT[63,64] and the University of Virginia,[65] and by the Cambridge Thermionic Corporation[66,67] and General Electric (US).[7] Interest in this technique revived in the late 1960s in Japan,[68] Israel[69,70] and the UK.[70–3] The variation of inductance of an electromagnet in the proximity of a ferromagnetic object depending on the separation between the two is utilised in this method to regulate the current and hence the force of attraction. This is achieved by incorporating the electromagnet within an LCR circuit tuned in such a way that when the object to be suspended moves away from the electromagnet the circuit tends to become resonant, thus increasing the current and hence the force acting on it (see Fig. 1.9). Conversely when the object moves towards the

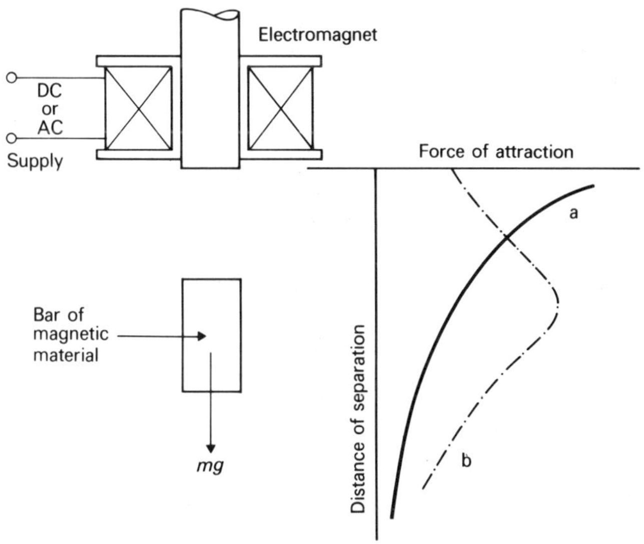

Fig. 1.9 Geometry and force–distance curves for a single magnet suspension scheme: (a) with DC excitation; (b) with AC excitation and series capacitor.

electromagnet the current and the force of attraction diminish. If, therefore, the force is balanced against that of gravity at some distance of separation it is possible to get a statically stable sit point for suspension of the body. However, the tuned LCR circuits possess large time constants, which means that once disturbed from this static stable point the object usually goes into a divergent oscillation unless some means are employed to control and speed up the current changes or to provide damping in some other way. Kaplan[73] found that at frequencies of the order of 6 to 26 kHz, leaky capacitors ranging from 0.4 μF to 0.02 μF provided adequate damping to obtain suspension of a ferrite disc and rod weighing 3.5 g and 13.5 g respectively (Fig. 1.10). In other methods, oil damping is used: the body to be suspended is submerged in oil.

The stiffness of suspension using the AC tuned circuit method tends to be rather low for many applications. The main disadvantages, however, stem from the fact that at the static sit point the circuit is predominantly inductive and hence reactive power input is rather large; and that the iron structure including the object to be suspended must be laminated. Thus although this method seems to offer at first sight an inherently stable force–distance characteristic[71] and therefore considerable advantages for suspension of ferromagnetic bodies, rather disappointingly it suffers from too many drawbacks to have resulted in many practical applications.

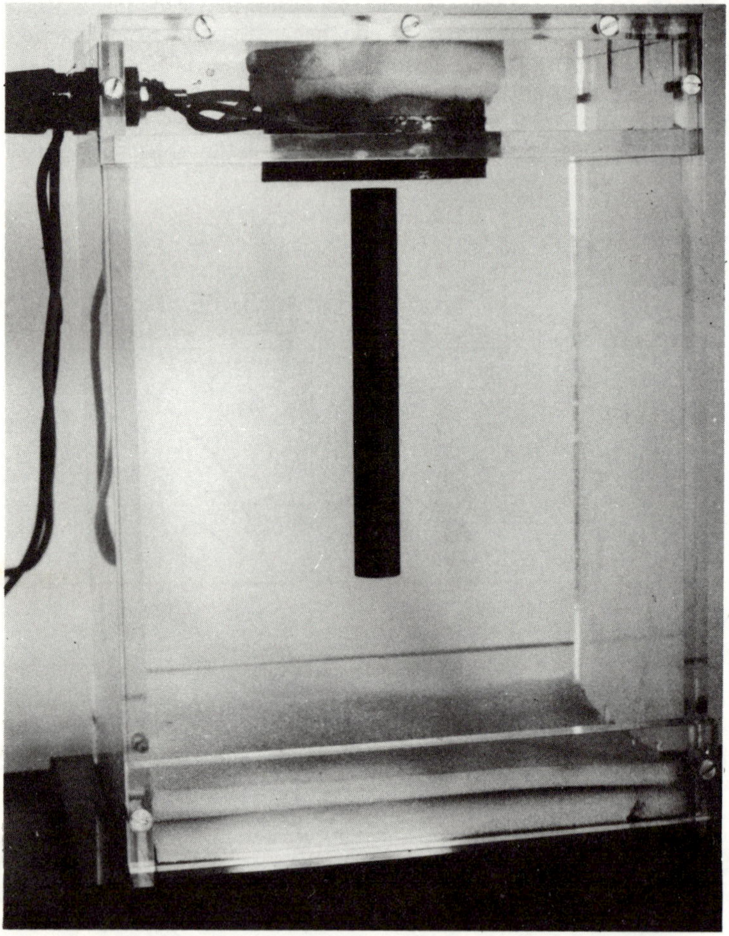

Fig. 1.10 A ferrite rod suspended using a tuned LCR circuit system.

1.2.8 Suspension using controlled DC electromagnets

The method is at present by far the most advanced technologically and is the subject of world-wide investigation for advanced ground transportation schemes and also for application in contactless bearings for high and very low speeds.

The first proposal for a controlled magnet attraction scheme appears to be by Graeminger[74] in 1911 for a vehicle suspended below an iron rail by a U-shaped electromagnet carried on the vehicle and facing the underside of the rail. A gap was to be maintained between the electromagnet and the rail by means of a mechanical or fluid pressure sensing device which would vary a resistance in series with the magnet winding or vary an airgap in the magnet core. As it stood the proposal did not have any practical potential. Anschüz-Kaempfe[75] in 1922 then suggested contactless centring of a floated sphere containing gyro motors using electromagnets. Position sensing was to be achieved by measurement of resistance of the conductive fluid between inner and outer spheres. Alternating current was also to be supplied to the support coils so that the eddy currents induced in the inner sphere would centre it by repulsion. However, the first amongst the present generation of suspension systems using active control of current in electromagnets is probably that of Kemper,[76,77] who proposed a vehicle suspended by electromagnets attracting it to the underside of a rail using either a capacitive or an inductive means of sensing distance below the rail. Part of the circuit also yielded a voltage proportional to the rate of change of the airgap for damping of the vertical oscillations. Kemper constructed a model consisting of an electromagnet with pole-faces of 30 x 15 cm and suspended a mass of 210 kg with an airgap flux density of 0.25 tesla. The airgap was 15 mm and the power consumption 270 watts. This remained the heaviest weight to be suspended using any method of electromagnetic suspension or levitation until the demonstration in 1979 of a 6.5 tonne vehicle by Messerschmitt (MBB) in Germany.

Much of the published work after that of Kemper on the development of the electromagnetic suspension schemes using controlled DC electromagnets and external position sensing was at the University of Virginia, particularly on rotor suspensions. The work carried out by Holmes[78] and Beams[79] in 1937 onwards was for rotors of high speed centrifuges required in the fields of biology and medicine, typical speeds being 77 000 rpm for a 3.97 mm diameter rotor. The other applications proposed were for testing bursting speeds of spheres such as ball bearings; testing adhesion of metal films; turbomolecular pumps for high vacuums free of bearings requiring lubricants; and magnetic suspension balances capable of recording weight changes of 5×10^{-11} g in a suspended weight of 2.3×10^{-6} g.

The same principle has been used to suspend aircraft models in wind tunnels[80,81] and appears to be the first instance of control of the three degrees of freedom of a suspended body. Since the objectives are to determine the forces acting on the aerodynamic model, the system is in effect a balance. Apart from the fact that it is virtually impossible to achieve an interference-free wakeflow field without a suspension system, the accuracy of such a scheme is more compatible with more recent requirements in aerodynamics. Further, magnetic suspension helps the investigation of more subtle aerodynamic details and improves techniques for studying aero-vehicle stability.[82,83] The importance of the method can be seen by the fact that all major aerodynamic research centres in the world have resorted to it at one time or another. Although this application appears to have originated in France[80,81] it was soon taken up by others: in the UK at the University of Southampton[84] and at the RAE;[85] and in the US at MIT-ARL,[86] AEDC,[87] University of Virginia,[88] Princeton University,[89] University of Michigan,[90] and NASA.[91]

There has been considerable activity since 1971 in the field of advanced ground transportation schemes using controlled DC electromagnetic suspensions, the first demonstration being that of the 6.5 tonne vehicle by Messerschmitt Bölkow Blohm (MBB) operating on a 700 m track (see Fig. 1.11). This was closely followed by another demonstration in Germany by Krauss-Maffei in 1972, at the University of Sussex[92] in 1974 (see Fig. 1.12), and at Japan Air Lines (Fig. 1.13) and General Motors (Fig. 1.14) in 1975.

In 1977 it was reported[93] that MBB had tested a rocket propelled vehicle Komet II at speeds in excess of 400 km h^{-1} (see Fig. 1.15).

Fig. 1.11 World's first full scale experimental vehicle using controlled DC electromagnets for suspension and linear induction motor for propulsion; demonstrated by Messerschmitt Bölkow Blohm, W. Germany, in 1971 (photograph courtesy of MBB).

Fig. 1.12 University of Sussex four passenger, 1 tonne vehicle using controlled DC electromagnets for suspension and linear induction motor propulsion.

Fig. 1.13 Japan Air Lines vehicle HSST-01 using controlled DC electromagnets for suspension and linear induction motors for propulsion (photograph courtesy of Japan Air Lines).

Fig. 1.14 Experimental vehicle using controlled DC electromagnets and linear induction motors; built by General Motors Research Laboratories at Warren, Michigan, USA (photograph courtesy of General Motors).

Fig. 1.15 MBB 8.1 tonne test vehicle KOMET, tested on a 1.3 km track at 401.3 km h^{-1} using water rockets for propulsion (photograph courtesy of MBB).

It is also reported that a 31.5 km track is to be constructed between Meppen and Papenburg in Emsland, Germany, and in June 1979 a 68 passenger, 36 tonne vehicle was demonstrated by MBB in the Hamburg exhibition. There are reports[94,95] that Japan Air Lines have demonstrated a 2.3 tonne, 7 m long coach for 9 passengers at Kawasaki, an industrial suburb of Tokyo, at speeds of 100 km h^{-1}. The Japan Air Lines programme is directed specifically at linking the Nerita and Haneda airports to Tokyo.

1.2.9 The mixed μ system of levitation

Earnshaw's[2] discussion of the stability of bodies in inverse square law fields showed that any equilibrium in such a force field is not stable. This result also applies to fixed charge or current distributions in free space but not to systems whose permeability differs from that of free space. Braunbeck's[3] extension of this theoretical analysis is that where the permittivity or permeability of the system is somewhat less than that of free space, and when dielectric or magnetic bodies are present, stability is possible. Bevir[96] has recently examined in more detail mixed systems where permeability in some places is less than that of free space and in others greater than that of free space, and has shown that in a few cases stable suspensions can be obtained.

A necessary, but not sufficient, condition for stability is div $\mathbf{F} < 0$, where $\mathbf{F}$ is the force vector acting on a body; Table 1.1[97] indicates how the sign of div $\mathbf{F}$ may depend only on the nature of the system and not on its geometry.

Table 1.1 indicates that there exists a class of mixed systems with materials both of $\mu > \mu_0$ and $\mu < \mu_0$, i.e. iron and superconductors as well as normal conductors with eddy currents or constant flux coils which should provide stable suspensions.

These theoretical predictions have been verified in three small-scale experiments:

Table 1.1

System	Permeability	div **F**	Stability
Normal conductors with constant current and orientation	$\mu = \mu_0$	$= 0$	Marginally stable
Constant current coils and iron	$\mu > \mu_0$	> 0	Unstable
Constant current coils with superconductors, eddy currents or constant flux coils	$\mu < \mu_0$	< 0	Stable operation possible
Mixed system of coils, iron and superconductors	$\mu > \mu_0$ and $\mu < \mu_0$	> 0 and < 0 in different regions	Stable operation possible

In the first, iron washers or nuts weighing 1.5 g were used; and in the second, iron cylinders 110 mm in diameter, 40 mm in length and weighing 100 kg. Both types of object are reported[98] to have been suspended stably. In the second experiment, the iron cylinder was held in suspension in a superconducting coil with an intervening screen, but although these results were impressive, the configuration does not lend itself easily to applications involving vehicles. The third experiment reported uses two superconducting coils in a persistent current mode operation and arranged at approximately Helmholtz spacing to give a uniform field between them. An iron cylinder is placed in the gap between the coils and the movement of this cylinder in any direction gives an auto-regulation of the current in the short-circuited coils such that the iron cylinder is returned to the central stable position. Whilst this geometry is more adaptable for vehicle suspension, no such results are available. Clearly more evidence is needed before any of the claims for the superiority of this scheme over others can be taken seriously.

1.3 A comparative assessment of the contending technologies

In this preliminary and cursory assessment it is appropriate to point out that whilst the popular media may appear to be interested only in high speed ground transportation applications, important contributions have been made in the field of contactless bearings and instrumentation, as will be apparent from the bibliography. In the fields of both high speed vehicles and bearings, the most advanced technology is that using controlled DC electromagnets. No mention has yet been made of a variation of this class of system which combines the lift and propulsion functions. It is suggested that this may be achieved by using (a) voltage controlled variable frequency linear induction motors;[99] (b) linear reluctance motors;[100] or (c) linear synchronous motors.[101] In (a) the DC windings and in (b) and (c) the AC windings are on the moving member. A 2 tonne vehicle combining functions of lift magnets with those of the linear induction motors was demonstrated in 1972 (see Fig. 1.16), but the results reported at that time did not appear to be very promising.

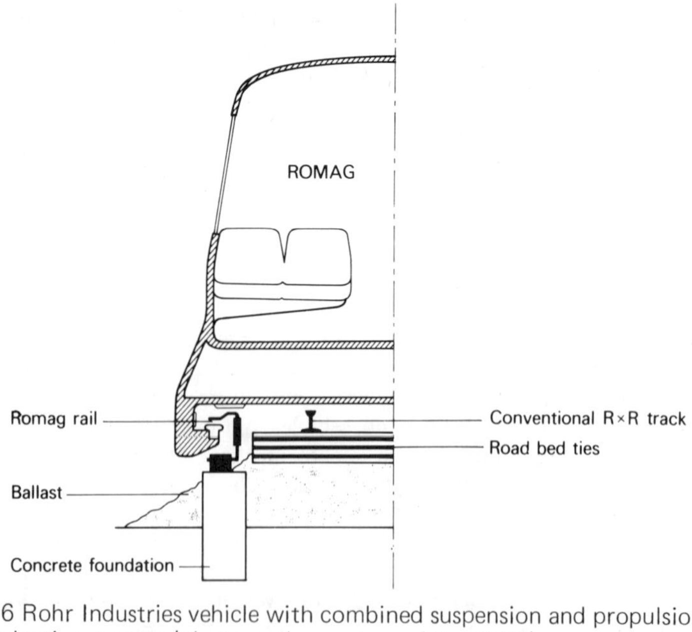

Fig. 1.16 Rohr Industries vehicle with combined suspension and propulsion using linear induction motors (photograph courtesy of the IEEE).

It is not yet possible to say whether synchronous motors will give better system configuration than separate lift magnets and propulsion units. However, in all these claims it is perhaps conveniently overlooked that the problems of propulsion units are by no means solved. Such problems concern their size and weight, and also the power electronics needed to drive them. Not a great deal is known about their dynamic characteristics, and a study of their interactions with the suspension systems is necessary. Although formidable, none of these problems is insoluble.

The most difficult problem as yet, with little evidence of satisfactory solutions by the protagonists of the advanced ground transportation schemes, is that of current collection at high speeds. A 100 passenger vehicle is likely to weigh about 30–50 tonne and when travelling at $450\,km\,h^{-1}$ will require power in excess of 3000 kW. The present state of the art of current collection does not indicate that this can be achieved. It remains to be seen whether or not the alternative of a wound track, where sections of track on which a vehicle is travelling may be switched, is cost prohibitive.

It must be borne in mind that whilst the contactless suspension technologies for passenger carrying vehicles are contending for transport systems of the future, so are conventional wheeled systems. It is not the purpose of this book to be pessimistic about contactless suspension technologies, nor is it the intention to be over-enthusiastic to the extent of ignoring unsolved problems and supporting unrealistic claims. The field of electromagnetic suspension is an exciting challenge to researchers in electrical and control engineering, academic and otherwise; and it is likely that this interest in research will continue for some time, with or without advanced ground transportation schemes. Although all the methods of suspension or levitation listed in Section 1.1 and described in Section 1.2 are interesting in their own right, it is considered that concentrating on the four methods with greatest potential for applicability is likely to be more useful. The four methods are:

(i) permanent magnets in repulsion;
(ii) eddy current repulsion using mains frequency excitation (magnetic river);
(iii) superconducting methods;
(iv) attraction using controlled DC electromagnets.

These four methods are discussed in the following chapters.

2

Levitation using permanent magnets

2.1 Introduction

During the 1970s a new class of materials for making permanent magnets was developed, based on cobalt and some rare earth elements. The improvement is so great that the cobalt rare earth magnets are in a class by themselves. In terms of their resistance to demagnetisation the new materials are 20 to 50 times superior to the previous best, alnico, and their magnetic energy is 2 to 6 times greater. Whilst there have been innumerable attempts to use the force of repulsion between permanent magnets in applications such as load relief in bearings,[1] the older materials have suffered from drawbacks of demagnetisation and possible reversal of one of the magnets in the case of a mismatch and relatively weak forces of repulsion. The new cobalt rare earth materials have now radically altered this picture. Like poles of such magnets can be made to approach without suffering more than a small percentage loss in magnetisation and even this loss is generally reversible. Such magnets can be brought together an infinite number of times, and after the first two or three times the force of repulsion does not change. A current carrying coil on one of these magnets can be energised in a direction such that the effective polarity of the combination is reversed, and yet when the current in the coil is removed the magnet remains virtually unaffected.

The design of magnets for repulsion devices or for applications involving combination with current carrying coils involves new ideas and new methods as well as some understanding of the magnetic properties of the new materials.

2.2 Review of properties of permanent magnets and magnetic materials

Ferromagnetic elements such as iron, cobalt and nickel have atoms in which one electron shell contains fewer than the maximum number of electrons. In such unfilled shells there are one or more unbalanced electron spins giving rise to a small magnetic moment and making the atom itself a tiny magnet. Ordinarily in a large collection of such atoms the atomic magnets point in various directions and cancel one another. If a sample of ferromagnetic material is placed in a magnetic field, however, the individual atomic magnets tend to line up so that when the sample is removed from the field, it retains a net residual magnetism. The total magnetisation, indicated by the symbol M, is the sum of the contributions of all the elementary atomic magnets. Magnetisation in any direction reaches its saturation value when all the atomic magnets are parallel and pointing in that direction.

Magnetic materials are traditionally divided into two categories, hard or soft, depending on the ease with which they may be magnetised and demagnetised. A hard material has a wide hysteresis loop and a soft material has a narrow one. Each reversal of a loop represents energy lost. Therefore soft materials are suitable for certain electrical devices such as transformers because the material is subjected to a reversal of magnetisation many times a second. Hard materials, on the other hand, have the properties required in the applications of permanent magnets, particularly those utilising the forces of repulsion such as floatation or levitation, since the width of the hysteresis loop determines the intrinsic coercive force. This width can range from less than 8 kA m^{-1} in the alloys used in telephone equipment to hundreds of kA m^{-1} in the new cobalt rare earth materials. For comparison the earth's magnetic field is a little under 40 A m^{-1}.

Magnetic materials are generally described by their hysteresis loops in which the vertical axis is the total flux density B rather than the magnetisation M. The total flux density includes the contributions of both magnetising field H and magnetisation M. In electrical engineering the performance of devices is more closely related to the total flux density. When considered in relation to the physics of magnetic materials, however, the magnetisation M is an indication of how the material acts in response to a magnetic field and thus both B and M are useful.

The upper left hand quadrant of a B–H loop is important in engineering when trying to determine the quality of permanent magnets. This is illustrated in Fig. 2.1.

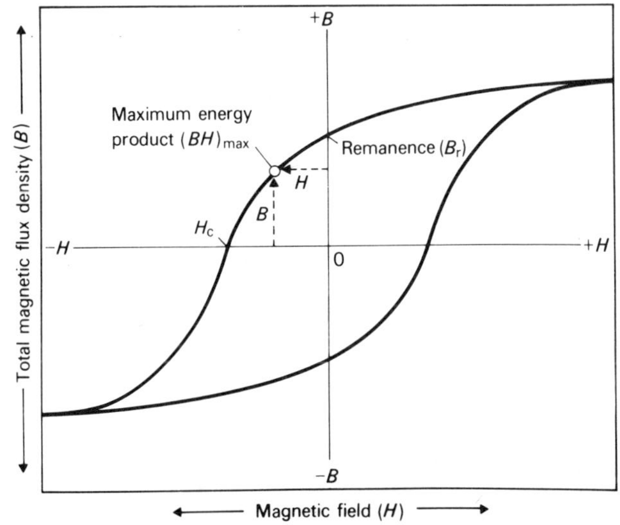

Fig. 2.1 B–H loop of the magnetic material of a permanent magnet showing the points of maximum energy product, remanence and coercivity. The maximum energy product $(BH)_{max}$ is used by engineers as a measure of the quality of the magnet.

Three points on this section of the curve are significant. First is the intersection with the vertical axis, i.e. the point where the magnetising field has been removed but the material retains its magnetisation. This value is termed the remanence, B_r. If the magnetising field is now reversed the value of the imposed field H_c which reduces the total flux density to zero is called the coercive force. Further, each point on the loop represents some value of B times H. The point where this value reaches a maximum is known as the maximum energy product $(BH)_{max}$. This maximum may be used as an index of quality for permanent magnets.

A field equal to the coercive force applied to alnico magnets on the one hand and to the new rare earth magnets on the other produces remarkably different results. Once the alnico magnet has been driven to zero flux density the flux rebounds only slightly. With the cobalt rare earth magnets, however, whilst it is possible to drive the total flux density to zero, as soon as the coercive force is removed the flux density rebounds almost to its original value. The reason for this is that the cobalt rare earths have values for intrinisic coercive force. As a consequence a demagnetising field can drive the total flux density to zero or even below zero without affecting the intrinsic magnetisation M of the material. This resistance to demagnetisation does not appear in the quantity maximum energy product but it is equally important in the context of these new magnetic materials. Cobalt rare earth materials have values of intrinsic coercive force from 20 to 50 times greater than those of conventional permanent magnets (see Fig. 2.2).

The property of the new rare earth magnets to resist demagnetisation makes it possible to envisage many new applications, and permanent magnet motors and

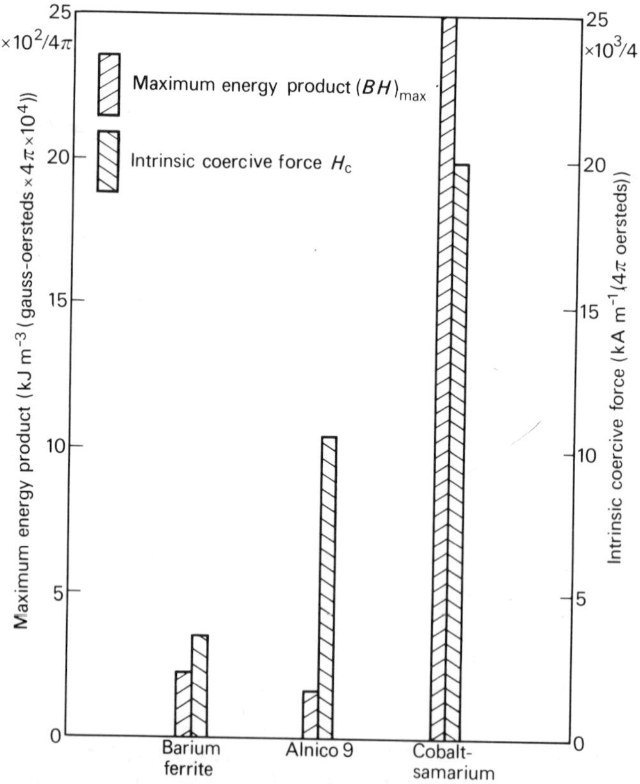

Fig. 22 Comparison of the intrinsic coercive force and maximum energy product for barium ferrite, alnico and cobalt-samarium.

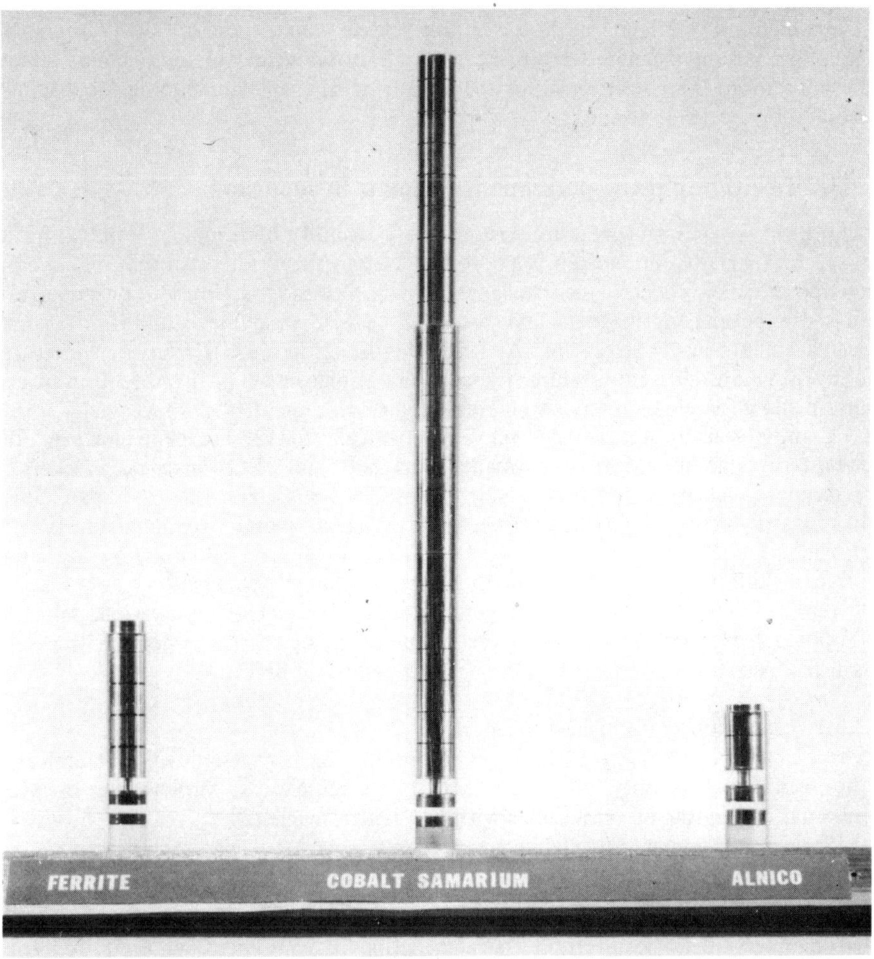

Fig. 2.3 Comparison of repulsive force of two discs of alnico, barium ferrite, and cobalt-samarium magnets. The weights that can be supported by the three materials when the gap is the same between the repelling magnetised discs are in the ratio 2:5:23 (photograph courtesy of Dr J. Becker, General Electric Corp., Schenectady, US).

alternators have already become accepted. In the context of the subject of this book repulsion forces between permanent magnets to support passenger carrying vehicles and trolleys capable of being moved about (say in factories) become technically feasible.

If gaps between opposing poles of disc magnets are compressed to the same extent it can be seen from Fig. 2.3 that the weights which can be supported by three magnetic materials, alnico, barium ferrite and cobalt-samarium, are in the ratio of 2:5:23. This means that if a 10 tonne weight or a vehicle is to be supported using ferrite magnets the total weight of the magnets (including those on the track as well as those on the levitated body) is estimated to be nearly 2 tonne.

The weight of the levitated body or the vehicle would therefore be 11 tonne. If cobalt-samarium magnets were to be used, the total weight of the magnets would be only about 0.5 tonne and thus the weight of the vehicle including the magnets would now be 10.25 tonne.

2.3 Levitation using permanent magnets in repulsion

Whilst there has been great interest in the possibility of complete levitation of a body by magnetic repulsion it is a corollary of Earnshaw's theorem that in a system composed solely of permanent magnets it is impossible for a body to be supported only by permanent magnets and for it to be held in stable equilibrium against displacements in all directions. At least one direction must remain for which the body is in unstable equilibrium; the smallest displacement in this direction brings into being a force which tends to increase the displacement.

Complete levitation without any mechanical contact has been achieved by means of a combination of permanent magnets and electromagnets. Backers[102] describes a magnetic journal bearing in which a shaft is supported with radial bearings of permanent magnet rings and in which support for axial location is provided by controlled electromagnets. There are several patents covering complete levitation systems. It is worth recording that a permanent magnet can be levitated above a saucer-shaped superconducting body. There are various schemes proposed for partial load relief of vertical bearings, or bearing systems with mechanical constraints, described by Geary[1] in Chapter 3 of his book.

2.3.1 Levitation of vehicles

The example cited with 250 kg of cobalt-samarium magnets added to it indicates the technical feasibility of passenger carrying vehicles, or trolleys for carrying materials, operating on tracks lines with permanent magnets.

Several advantages are claimed for vehicle systems using ceramic magnets in repulsion and electromagnetic or mechanical guidance systems. These are as follows:

a) no energy is required to achieve the lift forces;
b) lighter vehicles result from the elimination of equipment required for other means of support;
c) there is freedom in track design to suit various forms of transport, e.g. vehicles may be supported above or suspended below the track;
d) no maintenance is required;
e) there is no noise or vibration;
f) as the magnets are made of nonconducting materials there are no eddy currents and drag forces resulting from them;
g) economical and inherently safe operation follows from (a).

It is claimed that problems that could be caused by the attraction of strong metallic objects can easily be overcome by suitably designed or elevated track. It is also assumed that, if the need arises, such modern magnetic materials could be produced relatively cheaply on a large scale and the cost brought down to a low enough figure for their use to become economical.

Some of the advantages claimed above are applicable to most other contactless systems. The crucial factors are therefore whether the cost of a track of a permanent

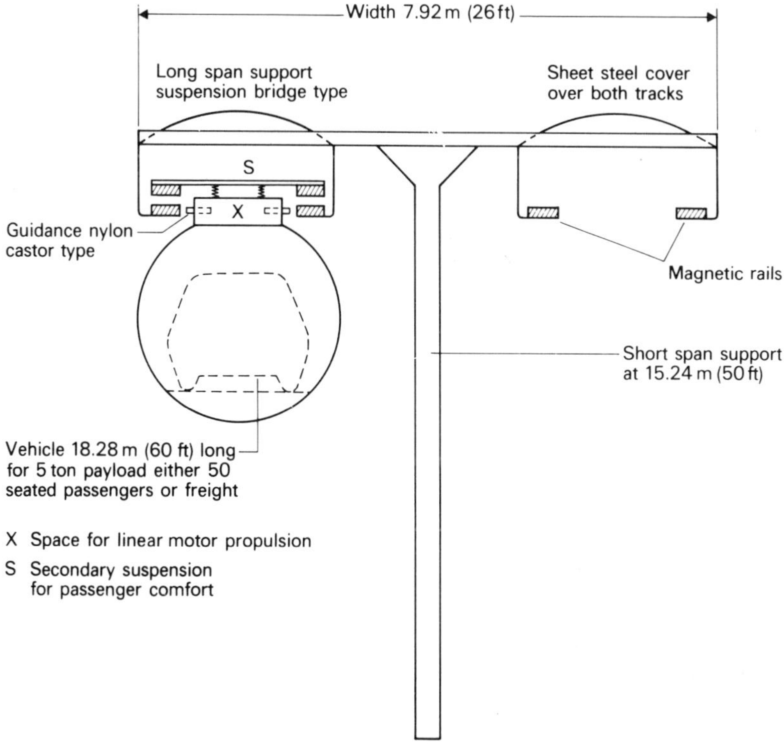

Fig. 2.4 Sectional view of overhead Magnerail trace as proposed by Polgreen.

magnet can ever be as cheap as that of steel rails. This is just conceivable for barium ferrite but barely so for cobalt-samarium and its descendents. The other factors are related to the provision of guidance forces and ride control. As is suggested, any schemes which use controlled electromagnets would inevitably detract from the advantages claimed above as well as having to use two different technologies.

The Westinghouse Company in the US was the pioneer in this field and was the first to suggest the use of ceramic magnets for transport applications. A one-passenger vehicle was developed at Westinghouse Research Laboratories[14] in the early 1960s. The main proponent of such schemes in this country has been Polgreen,[8–13] and Fig. 1.1 illustrates Polgreen's model. The proposed rail configuration is shown in Fig. 2.4. This arrangement of rails with a steel backing channel is also favoured by Bahmanyar and Ellison[18] because it is said to increase lift forces by better utilisation of the available flux and also by providing better mechanical protection for the brittle magnets.

2.4 Estimation of repulsion forces

2.4.1 Method based on experimental results

The calculation of forces of repulsion between magnets is difficult and, in general, even when scaling up using experimental results, may be subject to considerable errors. McCaig[103] has outlined a method based on experimental results for the configuration of Fig. 2.4 as well as a few others. The repulsion force per unit area for Ferroba III bar magnets is plotted against reduced distance, i.e. distance of separation divided by magnet thickness, in Fig. 2.5. The repulsion force is defined as the total supported weight per unit magnet face area with faces separated by a distance d at which the product (force × distance) is a maximum. The force increases with decreasing distance approximately such that it is doubled at $d/2.4$ and quadrupled at zero.

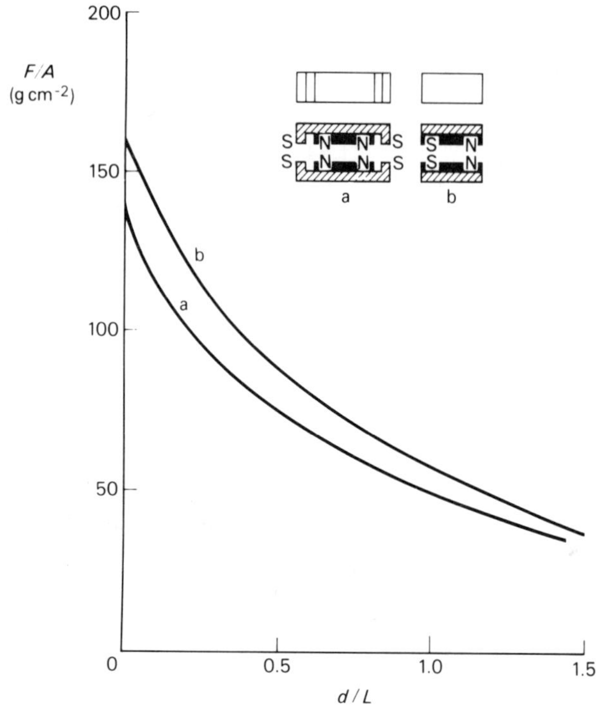

Fig. 2.5 Repulsion force per unit area for Ferroba III bar magnets plotted against reduced distance (i.e. distance of separation divided by magnet thickness).

Also given in Fig. 2.5 is a curve for another configuration b, and it can be seen that this would appear to be slightly better than the one considered previously. The results and measurements from McCaig's paper are tabulated in Table 2.1. The thickness L is the vertical thickness of the magnet without steel backing.

The force of repulsion depends on the magnet material, the arrangement of poles as described, the area, the shape and thickness of the magnets, the distance between the repelling pole surfaces and the dimensions of any steel parts used. Although these factors make the prediction of force difficult, if the same materials and configurations are used then correct use of the method of similitude should enable the forces to be estimated for other systems in which all the dimensions and distances

Table 2.1

Material	System	Area A (cm^2)	Thickness L (cm)	Distance d (cm)	Force/area $(F/A)_{opt}$ (g cm^{-2})	$(d/L)_{opt}$	Efficiency Fd/AL (g cm^{-2})
Ferroba III bars	a	3.02	0.48	0.67	36.7	1.4	52
Ferroba III bars	b	6.05	0.48	0.70	40.3	1.46	59

bear the same ratio to the one on which experimental results are available. The distribution of magnetisation and polarity in the magnets should then be the same. The force of repulsion should then vary as the superficial area of the magnets, i.e. as L^2 where L is any linear dimension. Since the volume and therefore the mass of the system varies as L^3, the force per unit mass varies as $L^2/L^3 = L^{-1}$. It thus appears that for any material and arrangement there is a maximum size above which the magnets cannot levitate their own weight. This argument is valid if the distance between the magnets is increased as their size increases. In practice the distance of separation is more or less constant and increased loads can be supported by an increased number of magnets.

In Table 2.1 and Fig. 2.5 the weight of the supported magnet is subtracted from the force of repulsion, but the weight of any steel parts is not. If, for a given pair of magnets, the force of repulsion F is measured for different distances of separation d, there is some value for d for which the product Fd is a maximum. This may be considered the optimum condition for the given arrangement. The quantities $(F/A)_{opt}$ and $(d/L)_{opt}$ follow from this notion of the optimal distance. Finally the quantity Fd/AL is a measure of the value of the product Fd per unit volume of the supported magnet.

The arrangements using mild steel pole pieces are very efficient, but in assessing the merits of different systems the weight of any mild steel in the support magnet must be taken into consideration. The cross section of the flux paths must also be sufficient not to cause losses due to saturation.

2.4.2 Analytical estimation of repulsion forces

The force of attraction between two magnets (or magnetised bodies) is given by the expression $F = B^2/2\mu_0$ per unit area. The flux density is assumed in this case to be uniform but when magnets are in repulsion this assumption is no longer valid. In fact if two identical magnets were held in close contact the flux density along the faces would be everywhere zero but the force would be maximum. In order to arrive at the correct estimate it is necessary to consider the influence of one magnet on the magnetisation of the other. This may be possible by considering each magnet to consist of a distribution of poles, dipoles or current loops. McCaig[17] has attempted an analysis of the forces of repulsion between small disc magnets by considering each to consist of a number of elementary charged layers. Alternatively the same result could be obtained by treating each magnet face as a charged surface. With those methods it is essential to evaluate the mean flux density or its gradient at various levels above the supporting magnet. It is necessary to use weighting functions to take account of the non-uniform field distribution in order to obtain satisfactory results. Bahmanyar[104] claims to have circumvented these difficulties. It is assumed that the magnetisation vector is everywhere perpendicular to the pole face and that the upper part of the magnetisation

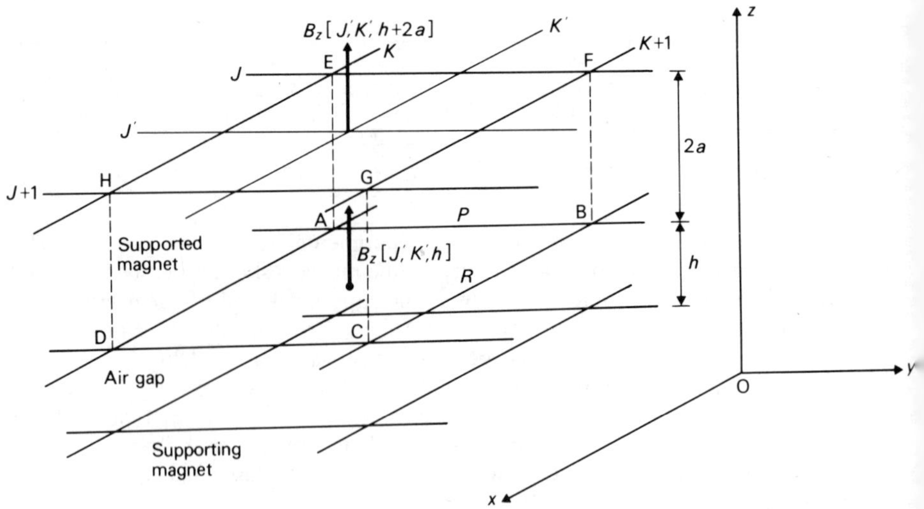

Fig. 2.6 Model used for repulsion force analysis by Bahmanyar using incremental mean flux density technique.

curve is parallel to the H-axis of the B–H graph. This is largely true for the modern high coercivity, large remanence, ceramic magnets. The first step is then to compute the normal (force-producing) component of the flux density B_z due to the supporting magnet, at every node of two $p \times q$ field matrices coincident with the two pole faces of the supported block. The values of p and q are made as large as can economically be handled by a computer. Two directly opposite elementary meshes ABCD and EFGH on the upper and lower faces of the supported block are considered bound by rows j and $j+1$ and columns k and $k+1$ of the field matrix. The mesh length along the x-axis (parallel to the direction of motion) is δR and along the y-axis δP. The gap length is assumed to be h and the magnet depth $2a$. Bahmanyar deduces in his thesis that B_z at a point height z above the supporting magnet is given by

$$B_z = \frac{M}{4\pi}\iint\left(\frac{z}{(x^2+y^2+z)^{3/2}} - \frac{z+2a}{(x^2+y^2+(z+2a)^2)^{3/2}}\right)\mathrm{d}x\,\mathrm{d}y \qquad (2.1)$$

and that the force of repulsion for this gap on the supported magnet is

$$F(h) = \frac{MPR}{\mu_0(p-1)(q-1)}\sum_{j'=1}^{p-1}\sum_{k'=1}^{q-1}[B_z(j',k',h) - B_z(j',k',h+2a)] \qquad (2.2)$$

where j' and k' refer to the rows and columns of the symmetrically superimposed repulsion force matrix having nodes at the centres of each elementary mesh, M is the intrinsic magnetisation and R and P are the dimensions of the magnet face along the x- and y-axes respectively. The last two terms in equation (2.2) are the mean flux densities over the elementary charged surfaces ABCD and EFGH respectively.

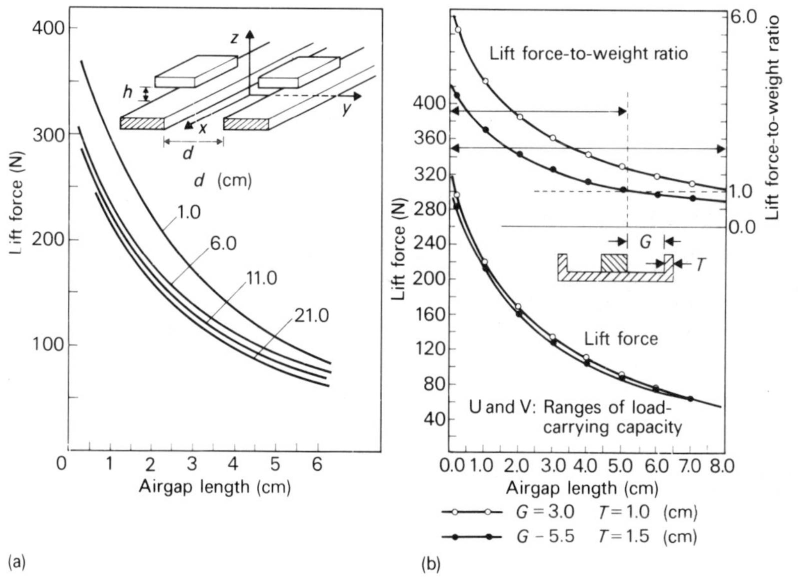

Fig. 2.7 Lift forces in permanent magnet repulsion schemes: (a) as functions of airgap length and rail spacing; (b) for steel backed magnetic rails and lift-to-weight ratio.

Bahmanyar has extended the analysis to magnetic track rail as well as the steel-backed structure similar to that proposed by Polgreen for Ferroba III. Results for these are given in Figs 2.7(a) and (b).

2.5 An assessment of the potential of permanent magnet technology for transport systems

There is little doubt that permanent magnets are now being made which can be used to support say 30 tonne passenger carrying vehicles. Two disadvantages of the system, i.e. necessity to provide lateral guidance and ride characteristics, were mentioned briefly in Section 2.3.1. These two problems together with the cost of magnets for several hundred km of track will now be considered.

The configuration of one magnet repelling another one above it is inherently unstable. In order to support vehicles it is proposed that there should be three or even four parallel rails formed of magnets and these will add considerable turning moments (roll) to the sideways displacements. There will therefore be the necessity to provide either mechanical guidance and restrictors or controlled DC electromagnets of the kind to be described in Chapter 5 to counter both these forces. The system is therefore either no longer contactless or will be required to use another technology which at first sight is not very desirable.

The question of ride comfort is probably more serious than researchers in this field appear to realise. The response to a step disturbance of a pair of magnets in repulsion is shown in Fig. 2.8. This is found to be very underdamped. Secondly the

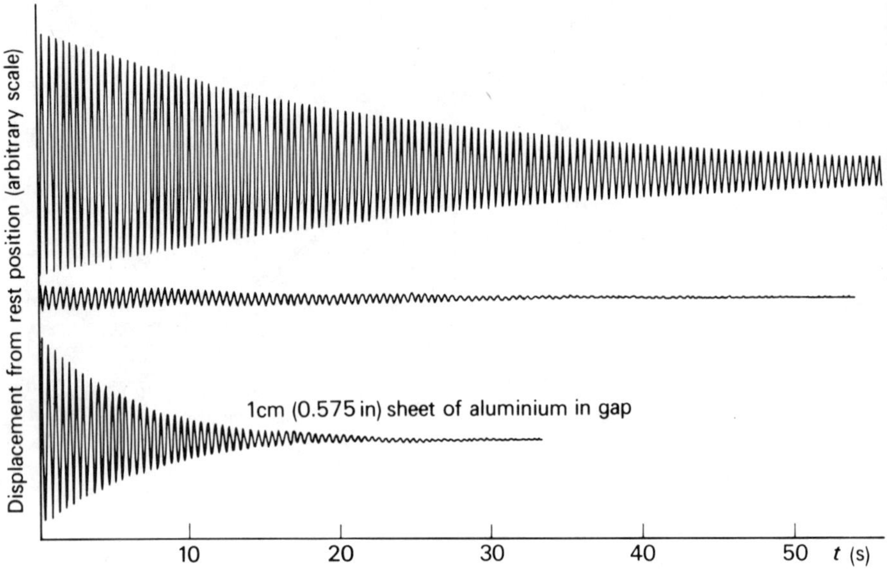

Fig. 2.8 Response of two permanent magnets in repulsion: two blocks of barium ferrite, 15 x 10 x 2.5 cm, at rest position 3.5 cm (1.5 in) gap.

stiffness of suspension for Ferroba magnets is about 10 kg cm^{-1}, even for the 1.0 to 2.0 cm gap range (Fig. 2.7), which appears to be rather low. A combination of these factors, unless controlled damping is provided, could be to produce undesirable ride characteristics similar to those on cars with spongy suspensions or, worse still, like a ship on rough seas.

The question of cost of track made up of permanent magnets must be preceded, even ten years after the first proposals for such a system by Polgreen and the Westinghouse Company, by a question as to whether sufficiently large quantities of these ceramic magnets would be available to build the tracks. Taking the example of the 10 tonne vehicle again, if it is assumed that such a vehicle is about 6 m long (the same as a single-decker bus) then the weight of magnets per metre run comes to approximately 40 kg. This means the weight of magnets on the track is of the order of 40 tonne per km (60 tonne per mile). The price of samarium-cobalt is high but assuming a cheaper version such as cerium/rare-earth/cobalt magnet to be available even at a quarter of this price, the cost of magnets alone is only a fraction of the total cost of the track. This has indeed been claimed by Polgreen since 1968. On the other hand, the problems of mounting these magnets on the track superstructure and those attracting rubbish of not inconsiderable weight are again likely to be more formidable and hazardous than the researchers claim them to be.

3

Levitation using superconducting magnets

3.1 Superconductivity and some properties of superconductors

A number of pure metals, and some alloys and metallic compounds, when cooled to temperatures approaching absolute zero (−273° C) exhibit two properties which are of interest to us here. The first one is that the resistance offered by the conductor to the flow of electric current vanishes; the material is therefore called a superconductor. Once an electric current has been set flowing around a closed loop of such a conductor the current will continue to flow and will produce a magnetic field. The loop thus carries a 'persistent' current and behaves like a permanent magnet. The second property is that superconductors in this state reject magnetic flux totally and thus behave as perfect diamagnetic materials with permeability[26] $\mu_r = 0$.

Although these two properties are the only ones of interest for purposes of levitation, it is interesting to note that superconductors are part of the category of materials at temperatures approaching 0 K which are called superfluids.[105] Two types of superfluids are known. Liquid helium is representative of one of them (boiling point 4.2 K) when cooled to 2.19 K, and the other is represented by the superconducting state of the electrons. In 1911 Kammerlingh Onnes, the Dutch physicist, discovered the latter phenomenon with mercury. Just as in a superconductor electric charge is transferred without a voltage difference between its ends, superfluid helium can pass with ease through extremely narrow capillaries or fine cracks which would be practically impassable for an ordinary fluid. Furthermore superfluid helium allows an extremely easy transfer of heat; this phenomenon is sometimes described as 'heat superconductivity'.

Kammerlingh Onnes in 1908 was the first person to liquefy helium and so to achieve temperatures of only a few degrees above absolute zero. He used the new technique to measure the electrical resistance of various metals at liquid helium temperatures and found in 1911 that, surprisingly, the resistance of mercury drops suddenly to an immeasurably small value when the temperature falls below a certain value. This value he determined to be about 4.2 K, which is known as the transition temperature of mercury. The resistance of mercury as a function of temperature is shown in Fig. 3.1. In the course of time a great many other metals and alloys have been found to possess this property.

Kammerlingh Onnes[106] further discovered that this phenomenon, which he called superconductivity, is destroyed when a sufficiently strong magnetic field is switched on. He observed a very sharp transition for a well-defined magnetic field strength when the field was oriented parallel to the axis of the conductor and when

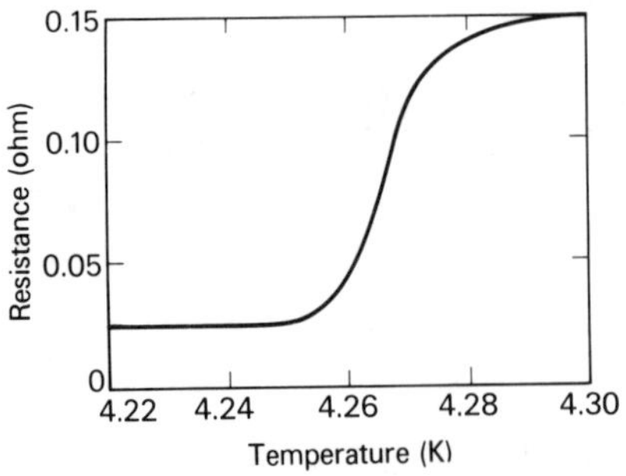

Fig. 3.1 The resistance of mercury as a function of temperature (as measured by Kammerlingh Onnes in 1911).

the currents used for measuring the resistance had very low values. The field necessary to destroy superconductivity, i.e. the critical or threshold field, depends on the temperature and disappears with a finite slope at the transition temperature. The curve representing the critical field H_c as a function of temperature, i.e. the threshold curve, has been measured for many superconductors (see Fig. 3.2).

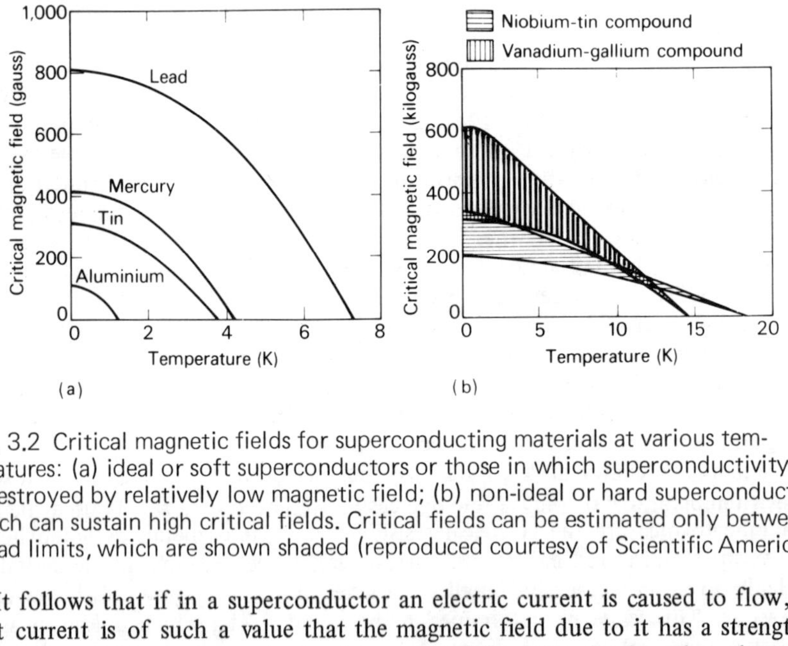

Fig. 3.2 Critical magnetic fields for superconducting materials at various temperatures: (a) ideal or soft superconductors or those in which superconductivity is destroyed by relatively low magnetic field; (b) non-ideal or hard superconductors which can sustain high critical fields. Critical fields can be estimated only between broad limits, which are shown shaded (reproduced courtesy of Scientific American).

It follows that if in a superconductor an electric current is caused to flow, and that current is of such a value that the magnetic field due to it has a strength at the surface of the superconductor in excess of the critical value, then the superconductive condition will be destroyed. Transition between normal and superconducting states is reversible, whether it is effected by changing the temperature or the ambient field strength.

In general there will be a magnetic field, however weak, surrounding a superconductor which is being cooled below the transition temperature. As the superconductive condition spreads through the material at transition, pockets of flux may in practice become isolated and cut off between the advancing fronts of the superconducting regions. The flux in these pockets will be compressed between the advancing fronts to the critical value of the flux density; then the surrounding superconductivity cannot get into the pockets and the flux cannot get out, i.e. the flux is 'frozen'. Thus it is difficult to obtain a superconducting body which is entirely free of magnetic moment although the magnitude of the magnetic moment is likely to be of significance in components for special applications only, such as levitated superconducting gyro rotors.

The materials discussed so far, which include many pure metals, are ideal or 'soft' superconductors. For these materials, transition and the exclusion of magnetic flux, apart from the frozen flux, is sudden and complete. Niobium (columbium) and lead, which fall in this category, have high critical temperatures of 8 K and 7.5 K respectively (see Fig. 3.2) but niobium also has a high value of critical field, about 320 kA m^{-1} (4000 oersteds). There are also non-ideal or 'hard' superconducting alloys and compounds which remain partly superconducting when penetrated by strong magnetic fields. Such substances are being developed, chiefly for use in the windings of superconducting electromagnets. Soft and hard superconductors differ in the manner in which they carry electric currents. In a soft superconductor, current is carried only in a thin surface layer. In hard superconductors, typical examples of which are compounds such as those consisting of vanadium and gallium (V_3Ga) or niobium and tin (Nb_3Sn) the current seems to be carried by filaments inside the material and the number of filaments can be increased by physically working the material. The effect of magnetic fields on typical soft superconductors is also different from the effect on hard superconductors. When the field is low, under 0.01 tesla (0.1 kilogauss), both types conduct current in a thin surface layer and the field is excluded from the interior of the material. A moderate field of 0.1 to 1 tesla (1 to 10 kilogauss) penetrates a soft superconductor, thereby destroying its superconductivity. In a hard superconductor, however, superconducting flow seems to be transferred to thin filaments and in certain hard superconductors filamentary flow persists, even in high fields of 1 to 10 tesla (10 to 100 kilogauss) and beyond, thus making them candidates for magnet windings of high field modern superconducting magnets.

The synthesis of niobium-tin (Nb_3Sn) was reported by Mathias in 1954[107] as a compound that becomes superconducting at 18 K, and this compound remains, even today, one of the materials with the highest transition temperature. It was pointed out by Kunzler[108] in 1961 that this compound satisfied the following three essential requirements of a material suitable for the construction of superconducting magnets.

1) The material must remain superconducting in a high magnetic field; niobium-tin was found to be superconducting in a magnetic field of 8.8 tesla (88 000 gauss).
2) The material must sustain a high current density in the high magnetic field; niobium-tin was shown to sustain a current density in excess of 100 000 A cm^{-2} at 8.8 tesla (88 000 gauss).
3) The material, even if refractory, must be capable of being fabricated into a magnet; niobium-tin is extremely brittle. However, niobium-tin powder is enclosed in a niobium tube surrounded by a monel metal jacket to act as an

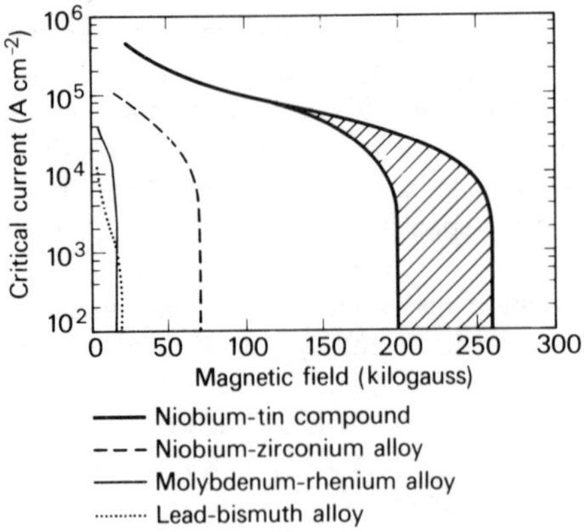

Fig. 3.3 Critical current density at varying magnetic field strengths. Placed in an increasing magnetic field the critical current density falls slowly at first but much more rapidly subsequently (reproduced courtesy of Scientific American).

isolator, and when drawn through successive reducing dies it is formed into a wire which is wound directly on a former. The niobium-tin powder is converted into the alloy Nb_3Sn by treating the coil to a temperature of 1000°C.

The properties of niobium-tin in comparison with other hard superconductors are illustrated in Fig. 3.3.

3.2 Principles of superconducting levitation

3.2.1 Levitation using the Meissner effect

In the survey of levitation using superconductors only a passing reference was made to levitation using the Meissner effect.[26] A considerable amount of effort has, however, been devoted to utilising this principle for levitating gyro rotors.

When a bar magnet is brought near a horizontal superconducting surface we may regard the flux as being compressed into the space between the magnet and the magnetically impermeable surface of the superconductor (see Fig. 3.4). Provided that the compression of the flux does not lead to the critical flux density being exceeded, the magnet will float on the superconducting surface in stable equilibrium in the vertical direction. If the surface is sufficiently large to be considered infinite in relation to the size of the magnet then the magnet will be in neutral equilibrium in the horizontal plane. It will in effect behave as if there is a mirror image of the magnet in the plane superconducting surface. When the size of the plane is reduced until the horizontal equilibrium becomes unstable, stability

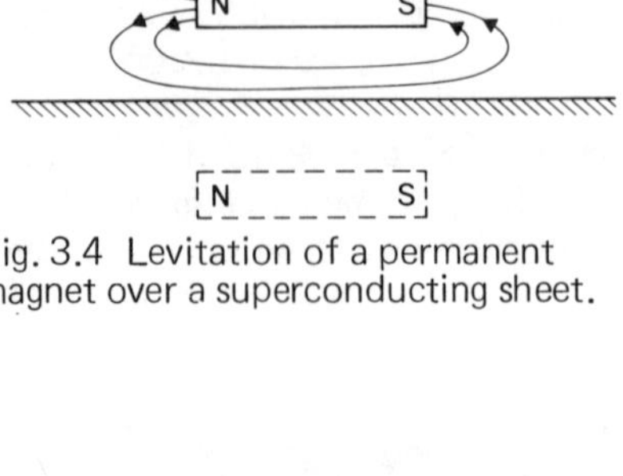

Fig. 3.4 Levitation of a permanent magnet over a superconducting sheet.

can be restored by dishing the surface. The experiment of levitating a bar magnet having dimensions of several millimetres above a saucer-shaped superconducting surface was performed by Arkadiev.[26,109] There are reports of other experiments[110–12] but Boerdijk[23,24] considered the possibility of levitating a magnet between an attracting magnet and a superconductor in the same way as he levitated a magnet between an attracting magnet and a piece of diamagnetic material.

Reported applications of magnetic levitation using superconductivity show that usually a superconducting object is levitated by the flux from one or more magnets. The levitated object is commonly a rotor. The magnets are usually superconducting rings or coils around which persistent currents flow. They may also instead be normal or superconducting electromagnets fed with constant current and, recently, may even be permanent magnets. Simon reports[27] theoretical and experimental investigations of the levitation of a superconducting hollow lead foil sphere, 1.25 cm in diameter and weighing 0.70 g, over a pair of rings or coils carrying currents in opposite directions or a pair of ring permanent magnets with their axes vertical. Although Simon concluded that levitation over a single ring was not possible this was subsequently disproved by Harding and Tuffias.[113] This work, following the investigations of Culver and Davis,[28] was directed towards the development of the 'cryogenic magnetic suspension gyroscope'. Harding and Tuffias report the analyses of different configurations of superconducting current carrying rings. They also describe other practical aspects such as the method for starting persistent current in the superconducting ring and the spot welding of niobium wire for making the joints in the ring. Rotors were solid niobium spheres or lead spheres plated by electrodeposition or vacuum deposition and made of sapphire, plastic, aluminium or magnesium. In the experiments described, a 3.8 cm diameter lead coated sphere weighing 8 g was levitated above a single ring carrying 1000 ampere turns. The largest sphere levitated was 4.25 cm in diameter and was made of solid niobium, weighing about 300 g. Buchold[114] also reports work on cryogenic rotor gyros at General Electric (US). Construction is described of a cryogenic gyro with a niobium spherical rotor levitated between 10 coils of niobium wire carrying persistent currents. The rotors are typically 5 cm (2 in) in diameter with a wall thickness of 1 mm (0.04 in) and weighing about 110 g. An equatorial rim constitutes the armature of a cryogenic electric motor as the means by which the rotor is driven up to speeds of 40 000 to 50 000 rpm. The rotor has a radial clearance from its housing of 0.35 mm (0.012 in). The interior of the housing is evacuated and vent holes provided in the rotor wall. Other factors, such as sensing of rotor speed and attitude and refrigeration are also described. There is a profusion of literature on the cryogenic gyros and other devices emanating around 1960 from General Electric (US).[115–19]

In Buchold's design of cryogenic bearings and the gyro,[114] the support of the rotating component is derived exclusively from the repulsion between a superconducting surface and a superconducting ring or coil carrying persistent currents and placed near the surface (see Fig. 3.5). Buchold found that in the arrangement of Fig. 3.5(a), where the shaft is coaxial with the rings, there is insufficient centring force on the shaft, i.e. the radial stiffness is low. The arrangement of Fig. 3.5(b) represents the configuration of his spherical gyro rotor. Here the coil axes are directed towards the centre of the sphere. There are advantages claimed for the rings to be persistent current carrying instead of constant and supplied by an

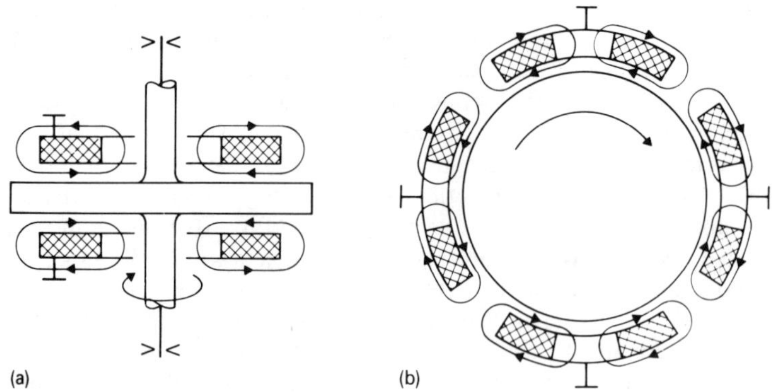

Fig. 3.5 Superconducting levitation for bearings: (a) axial levitation of a superconducting disc located between two fixed coils or current carrying rings; (b) radial levitation of a superconducting shaft or sphere located between fixed coils or rings facing its surface.

external source. In the former case, the total flux in the gap remains constant and therefore when the gap is reduced, the flux density and the force increase. In fact the force increases inversely as the square of the gap length.

Damping of any oscillatory motion of a superconducting body levitated over a current carrying superconductor is effected by locating a non-superconducting metal, such as iron or copper, near the current carrying member. Changes in the distribution of flux around the superconducting ring or coil due to the approach and recession of the suspended body produce eddy currents in the non-superconducting metal and thus provide a means of dissipating energy.

3.2.2 Levitation using eddy currents induced by motion of superconducting magnets

This approach to magnetic levitation is based on the repulsion induced by a magnet travelling above a conducting surface. Conventional electromagnets and even permanent magnets are capable of induced levitation at small clearances but they are incapable of producing magnetic fields of the size and intensity required for anything heavy, such as, for example, a passenger carrying vehicle. Superconducting magnets have, however, made this possible.

The phenomenon of electrodynamic levitation, as this effect is called, can be demonstrated very simply by spinning an aluminium drum at a sufficiently high speed and placing above it a suitably (curved) shaped magnet attached to a flexible cantilever. The principle of this method of levitation is depicted diagrammatically in Fig. 3.6. A current carrying loop is assumed to be moving past a somewhat larger short circuited loop. As the magnetic flux from the travelling loop begins to link the stationary loop, it induces an emf resulting in a current circulating in the loop and tends to oppose the change in the magnetic flux linking the stationary loop. The induced voltage is proportional to $d\phi/dt$, i.e. to the rate at which magnetic flux through the loop is changing, and the current is determined by the inductance and resistance of the loop. When the moving loop is directly over the stationary loop the flux linking does not change and the induced emf falls

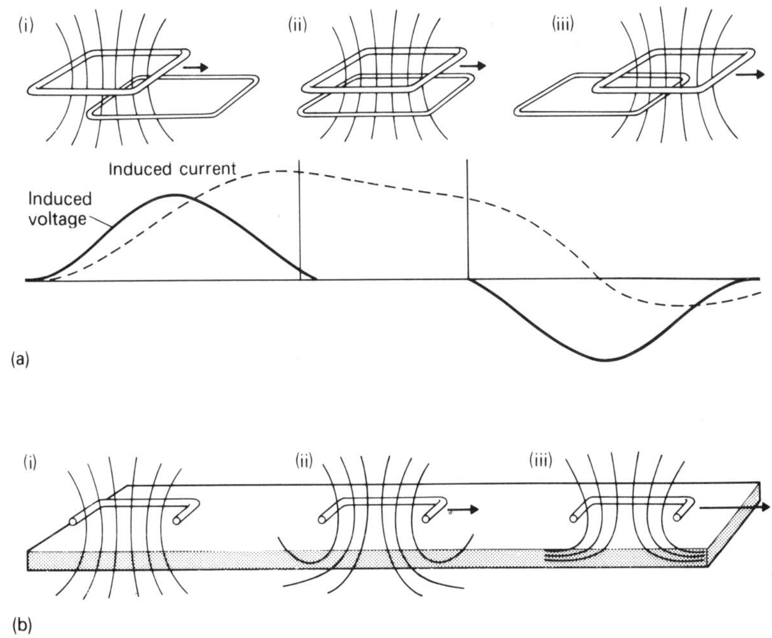

Fig. 3.6 Principles of repulsion levitation. (a) The loops above are supposed to be vehicle mounted, the ones below track mounted. In position (i) the increasing flux linkage induces an opposing current in the stationary loop. The induced current is zero in position (ii) and in the opposite direction in position (iii). (b) The guideway loops are replaced by a continuous sheet of conducting material. At high speeds the flux will barely penetrate the sheet and the situation almost resembles that in Fig. 3.4, the image in this case being a current loop carrying opposite current (reproduced courtesy of Scientific American).

to zero. The induced circulating current begins to decay owing to resistive loss in the stationary loop. As the travelling loop begins to move away from the stationary loop, the flux linkage begins to decrease and the induced emf is now of a reversed polarity. This brings the circulating current to zero and also produces a current in the opposite direction that is equal to the amount by which the originally induced current has decayed. This residual current persists for a fraction of a second after the travelling loop (or the vehicle carrying this loop) passes, and is called eddy current wake.

Both lift and drag are produced by the interaction of the two loops. Lift results from the fact that the induced current in the stationary loop flows in a direction opposite to the current in the approaching loop, and therefore repels it. This repulsion alone would not produce a drag, as its mechanical equivalent is a wheel passing over a stone. All the energy imparted in raising the wheel is regained when the wheel descends to its original level. Unfortunately, in the electromagnetic case, the current in the stationary loop decays as the travelling loop passes and, therefore, not all the approach energy is recouped in the departure. The energy lost is therefore equal to the resistive dissipation in the stationary loop. The opposite current induced in the stationary loop due to decay is attractive and therefore causes drag. An unusual phenomenon is that the faster the vehicle carrying the current loops (which may be superconducting) travels, the less is the drag. Decrease of drag with increasing velocity is rare in any transportation system.

All the early investigators thought in terms of a guideway composed of discrete loops or coils and there may in fact be an advantage in such arrangements. However, the loops can be replaced by continuous conducting sheets without any fundamental change in the interactions. So long as the coil or the magnet is stationary the magnetic field is unaffected by a conducting but non-magnetic sheet. As the current carrying coil begins to move, eddy currents, which will be distributed in nature, are induced in the conducting sheet. Their overall affect is to generate a magnetic field which opposes that of the moving loop and to keep it from penetrating through the conducting sheet, in the process giving rise to lift forces.

The induced eddy currents do not become strong enough to oppose the penetration of magnetic field through the sheet and to keep it out entirely until the current loop has reached a sufficiently high speed. Eventually, if the current loop is driven at an increasing speed, when it reaches the appropriate speed virtually none of the magnetic flux is allowed to penetrate. The conducting sheet now behaves as a magnetic shield. The moving current loop sees an image loop of opposite polarity, equidistant from the surface of the conducting sheet and thus repelling it. The force of repulsion becomes large as the moving loop approaches its image.

The main characteristics of electromagnetic levitation can be explained using these basic principles. If a current carrying coil accelerates at a constant height above a conducting guideway, it experiences a lift force at first proportional to the square of velocity but eventually reaching a limiting value. The drag force is at first proportional to velocity but then passes through a peak and decreases inversely as the velocity. At high speeds the drag decreases more slowly because of the skin effect: the induced eddy currents are largely confined to a thin layer near the surface of the conducting sheet and consequently the guideway appears to have higher resistivity. Unlike aerodynamic drag, the electromagnetic drag decreases with increasing speed and this is a remarkable feature of electrodynamic levitation. The lift force is proportional to the product of the perpendicular and parallel components (relative to the guideway) of the magnetic field and the drag is proportional to the square of the component of the magnetic field perpendicular to the surface of the guideway. The lift-to-drag ratio is therefore proportional to the ratio of the parallel component to the perpendicular component of the magnetic field. This relationship gives some insight into the way the configuration of the magnets on a vehicle can be optimised.

An interesting but very rough guide (which completely ignores the problem of overcoming the drag force) is that, if the magnetic field at the surface of the guideway is about 2.0 tesla (20 kilogauss), sufficient force will be generated just to support the magnet and its notional payload at speeds as low as 32 $\mathrm{km\,h^{-1}}$ (20 mph) and the limit of the lift force will be very nearly reached at 100 $\mathrm{km\,h^{-1}}$ (60 mph). The lift force is nearly 400 000 $\mathrm{N\,m^{-2}}$ (60 $\mathrm{lb\,in^{-2}}$), the same as the pressure in the tyres of a bus. It is therefore suggested that magnets about the size of the footprint area of a bus wheel tyre could levitate a bus, a suggestion which inspired the early investigators of a wheelless train.

3.3 Some design considerations of passenger carrying vehicles with cryogenic magnets

The evaluation and progress of the electrodynamic system of levitation for passenger carrying vehicles is linked, besides the configuration and geometry of the superconducting magnets, to a few other factors. The principal ones are the problems of suitable propulsion schemes. Although the levitation schemes are capable of operating at large airgaps, the linear induction motor represents a substantial fraction of the overall vehicle weight and also requires small airgaps in comparison. In addition, the peak of the drag force occurs at speeds of approximately 16 km h^{-1} (10 mph) and the thrust required to move past this peak is prohibitively high, whilst at the upper end of the speed range collection of power of the order of a few megawatts through sliding contacts would present some formidable problems. Secondly the ride quality analysis indicates that a guideway surface equivalent to a poor road surface can be tolerated with the high clearance suspension, but without some form of active damping the inherently oscillatory nature of the suspension will fail to satisfy the passenger comfort criteria. Both the lift mode and the lateral guidance need to be damped. The problem of lateral guidance must be incorporated into the guideway design.

This discussion has precluded any economic or environmental considerations of a new guideway alignment, such as that envisaged for any advanced ground transportation scheme. These are perhaps of even greater importance than the technical considerations.

3.3.1 Repulsion and drag forces of the superconducting levitation schemes

A current carrying coil moving with a vehicle induces eddy currents in a conducting roadbed or guideway and these eddy currents interact with the coil in such a way as to repel and support it. There is no lift force at zero speed so the vehicle must be supported in some other way at low speeds and the commonly suggested solution is wheels. As the vehicle speed increases, for a magnet at constant height, the lift force F_L increases rapidly at first and then levels off approaching the image force F_I at high speeds;[32] F_I is the force between the coil and its image in the guideway. The image magnet is an identical magnet placed an equal distance below the surface of the guideway and carries equal and opposite current. As explained in Section 3.2.2, in addition to the lift force there are the drag forces F_D on the magnet as a consequence of the eddy currents or Joule losses in the conducting guideway. The value of F_D passes through a peak at a speed of 16 km h^{-1} (10 mph) and then drops off continuously as the speed is increased. The relatively low value of the magnetic drag at high speeds will add to the efficiency of this suspension.

The important parameters which characterise the suspension are lift and guidance forces per unit magnet weight and the ratio F_L/F_D. Leaving aside the problem of guidance for the time being, calculations have been carried out[120,121] for F_L and F_D on rectangular coils moving with a velocity v above a conducting plate of arbitrary thickness and infinite extent. An analytical fit to the calculations, if the plate thickness $T < \delta$ where δ is the skin depth of penetration and the permeability of the plate material $\mu = \mu_0$, i.e. non-magnetic, is

$$F_L = F_I[1-(1-v^2/\omega^2)^{-n}] \tag{3.1}$$

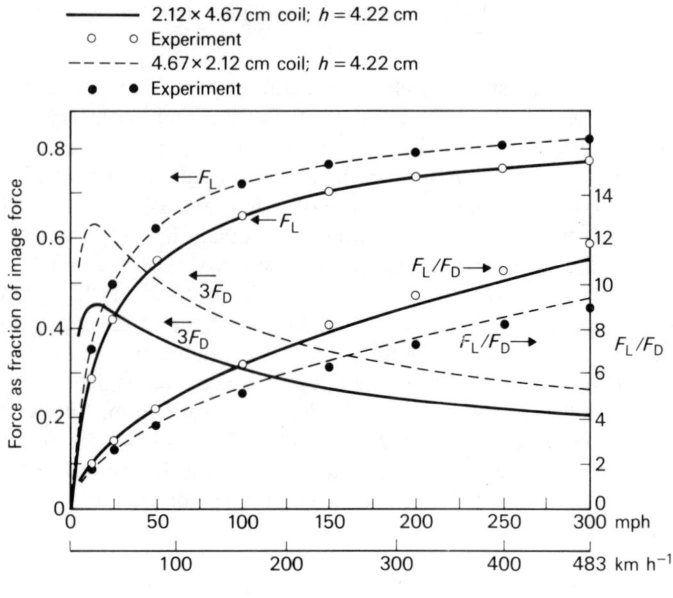

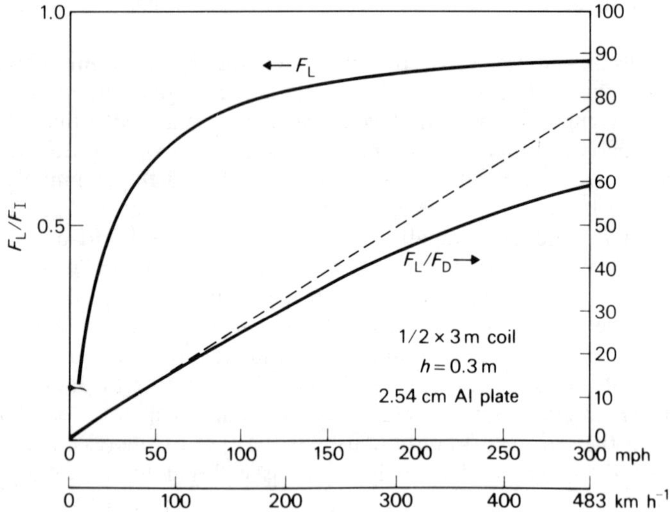

Fig. 3.7 Lift and drag forces on rectangular coils above aluminium guideway. (a) Lift force, drag force and lift-to-drag force ratio as a function of speed. Second dimension of the coil in the direction of motion. Infinitely thick aluminium slab. (b) Lift force, and lift-to-drag force ratio as a function of speed of coil over 2.54 cm thick aluminium plate. Motion of coil in the direction of the second dimension of the coil (reproduced courtesy of the IEEE).

and[122,123]

$$F_D = (\omega/v)F_L \quad (3.2)$$

where $$\omega = 2/\mu_0 \sigma T \quad (3.3)$$

The quantity ω has the dimensions of velocity and σ is the conductivity. The number n is determined by the dimensions of the coil. Equation (3.2) is an exact result. For the coil of Fig. 3.8(b), with dimensions 0.5 X 3 m (3 m side parallel to v), n is approximately equal to 0.2 and for an aluminium plate 1 in (2.5 cm) thick, $\omega = 6300\,\mathrm{m\,h^{-1}}$ (3.9 mph).

If the conducting plate is thick, eddy currents are limited by skin depth δ and in equation (3.3) the value of ω is obtained by substituting δ for T, where $\delta = \lambda_{\mathrm{eff}}/(\pi\mu_0\sigma v)^{1/2}$ and λ_{eff} is the effective wavelength for the geometry under consideration.[123] For a wide rectangular coil moving perpendicular to its length at a height h small compared to its width, $\lambda_{\mathrm{eff}} = 16h$. For a narrow long coil of long side $2b$ parallel to v, $\lambda_{\mathrm{eff}} = \pi^2 b/2$. For plates of intermediate thickness, F_{L} and F_{D} have been determined numerically using Fourier methods.[121] In general it is found that increasing the levitation height and the length of the coil in the direction of motion of the coil improves the ratio $F_{\mathrm{L}}/F_{\mathrm{D}}$ but increasing the levitation height beyond 30 cm is not considered practical as the lift force diminishes with increase in the levitation height. Fig. 3.7(a) shows calculated and experimental values[121,124] of F_{L} and $F_{\mathrm{L}}/F_{\mathrm{D}}$ for two coils 2.12 x 4.67 cm at a suspension height of 4.22 cm over a thick aluminium plate. The experimental results were obtained by suspending the superconducting coils over the rim of a rotating aluminium wheel 61 cm in diameter and 15.2 cm wide. The lift force on a 0.5 x 3 m coil at a suspension height of 0.3 m, together with the ratio $F_{\mathrm{L}}/F_{\mathrm{D}}$, is shown in Fig. 3.7(b). The drag force is shown in Fig. 3.8(a). Four such coils, each carrying 3.6×10^5 ampere turns, would support a 50 tonne vehicle travelling at $v = 483\,\mathrm{km\,h^{-1}}$.

The guideway plate acts very much like a 'thin' plate up to approximately $160\,\mathrm{km\,h^{-1}}$ (100 mph). In this range it can be seen that $F_{\mathrm{L}}/F_{\mathrm{D}}$ is almost linearly proportional to speed and is essentially the same as that given by the thin plate analysis (dashed curve in Fig. 3.7(b)). The eddy current distribution above $160\,\mathrm{km\,h^{-1}}$ is not uniform but is now limited by considerations of skin depth δ which is approximately 2 cm at $483\,\mathrm{km\,h^{-1}}$. The difference between the value of $F_{\mathrm{L}}/F_{\mathrm{D}}$ predicted by thin plate theory and the actual value is seen in Fig. 3.7(b) to be almost 25% at $483\,\mathrm{km\,h^{-1}}$. Whilst the theoretical calculations are based on a rectangular geometry for the coil, in practice the corners will be rounded. However, this does not have a significant effect on either F_{L} or F_{D}.

There is in general a transverse force on the magnet if the guideway conducting plate is of finite width and the force is such as to push the magnet off the track. Calculations and experiments show[125] that if the track width exceeds the magnet width by about twice the levitation height the degradation in lift from that over an infinite plate is very small. The results from a 0.5 x 3 m coil for lift force F_{L} and transverse force F_{T} for height above conducting plate $h = 0.3$ m are shown[124] in Fig. 3.8(b). If the distance h' of the coil from the edge of the plate exceeds the suspension gap h, F_{T} is small and F_{L} is essentially the same as for an infinite plate. Since the drag force scales roughly with the sum of F_{L} and F_{T} the lift-to-drag ratio is degraded by about 10% if the track width exceeds the magnet width by only $2h$.

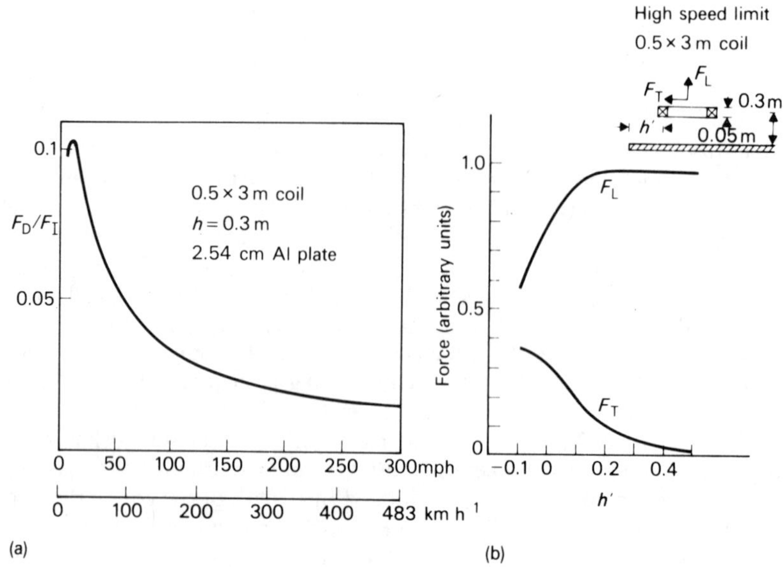

Fig. 3.8 Lift, drag and transverse forces on coils over aluminium plates: (a) drag force as a function of speed; (b) lift and transverse forces on a coil near the edge of the conducting plates (reproduced courtesy of the IEEE).

It is necessary to have additional levitation surfaces in the guideway in order to provide transverse guidance forces for a suspended vehicle. The surfaces are preferably vertical ones and the vehicle can either have separate guidance magnets to operate against these surfaces or the main levitation magnets can operate in a dual role and provide guidance against vertical sections of the guideway. The guidance force generated by the lift magnet constituted by the same 0.5 × 3 m coil at a levitation height of 0.3 m is shown in Fig. 3.9. In either case the guidance force

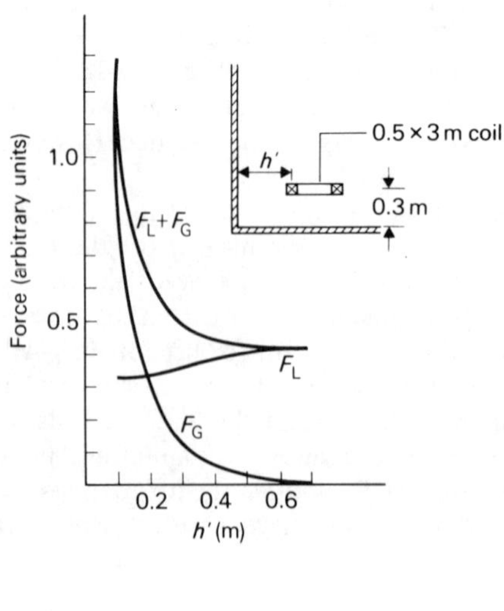

Fig. 3.9 Lift and guidance force on a coil near a conducting plate shaped as a right angled corner (reproduced courtesy of the IEEE).

produces extra drag which degrades the overall suspension performance criterion given by F_L/F_D because[122,124]

$$F_D = \frac{v}{\omega}(F_L + F_G) \tag{3.4}$$

where F_G is the guidance force.

3.3.2 Cryogenic magnet design and cryogenic magnet requirement

The design of a magnetically levitated vehicle utilising superconducting magnets is crucially dependent upon a good magnet design which will transmit forces safely from the superconducting magnet element, keeping the heat losses low and yet with minimum distance between the bottom of the dewar (vacuum jacketed vessel) and the superconducting wires.

The basic features of a cryogenic system for a vehicle operating on a U-channel guideway are shown in Fig. 3.10. There are eight magnets, four for lift and four for guidance; this is not necessarily an optimum number. The lift and guidance pair are shown in Fig. 3.10 at one of the corners and the arrangement allows a certain degree of redundancy and hence safety to be built into the system. The fringing

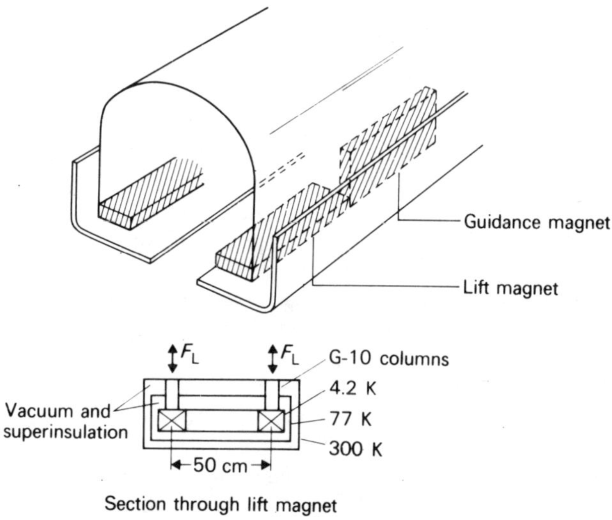

Fig. 3.10 Possible configurations of lift and guidance magnets in a U-shaped channel.

field of the lift magnet can provide guidance in case of failure of the guidance magnet and vice versa. In the example of the 0.5 x 3 m coil with 3.6×10^5 ampere turns the magnet and dewar assembly would weigh approximately 600 kg and hence give a lift-to-weight ratio of nearly 20. In order to transmit the lift and guidance forces from the magnet at 4 K to the frame at 300 K some composite material such as epoxy fibre glass with very low specific conductivity but high tensile as well as compressive and fatigue strength must be used. Using layers of metallic reflectors to keep the radiation losses to a minimum, alternating with insulating spacers, the conduction and the radiant heat energy transmitted from the magnet at 4 K to 77 K is of the order of 1 W for each.

A major source of heat leak into the cryostat is through the leads carrying the current in and out of the wires. For a well designed pair of leads this would, approximately, be less than 2 W per 1000 A pair since the vehicle magnets are by no means large by present superconducting magnet standards. If the magnets are operated in the persistent mode these losses would be even further reduced. Typical figures of heat load to the 4 K cryostat would be less than 3 W per magnet or less than 24 W per vehicle. There would also be AC losses in the superconducting windings due to the oscillatory motion of the magnets of the order of 1 W to 10 W per magnet for the typical 10–20 μm diameter filaments in a multifilament niobium-titanium composite.[123,126] The 4 K refrigeration for these magnets could come from on-board light weight cryogenic refrigerators although such units have not yet been fully developed. The 77 K or intermediate temperature refrigeration could be achieved by either liquid nitrogen, the heat capacity in the boiled-off helium or an intermediate temperature point on the on-board refrigerator although this would increase the size of the refrigerator.

3.3.3 Propulsion, acceleration and braking

The evolution of all the electromagnetic suspension or levitation schemes for passenger carrying vehicles is inextricably linked to the problems of propulsion There are two candicates being studied for these applications: one is the linear induction motor (LIM), the other is the linear synchronous motor (LSM). Linear induction motors have received serious attention for many years quite independently of magnetic levitation.[125,127–30] Among other things they have been used to transport liquid metal and metallic objects in conveyor lines. In transportation they were viewed as the only alternative to jet propulsion for tracked aircushion vehicles. Even earlier (and currently in one specific application) they were seen as a means of propelling wheeled vehicles above the speeds at which wheel traction fails.

Propulsion by linear induction motors involves some serious problems. Firstly, it requires that several megawatts of power be picked up from a live rail, a very costly operation that has yet to be solved, by vehicles moving at speeds in excess of 400 km h^{-1}. The alternative of generating power on board the vehicle would be difficult to implement if environmental considerations are to be met. It implies the installation of the equivalent of a 20 000 kW jet engine or gas turbine engine with attendant noise and air pollution problems. The alternative of a fixed stator winding along the track is at first sight not viable economically, although it was used on the Hamburg exhibition vehicle TR05 and is being proposed for the current German development in Emsland. Secondly, to be reasonably efficient a linear induction motor must operate at small gaps and at very nearly synchronous speeds. The latter requirement means frequency conversion equipment which until recently has tended to be heavy and bulky as well as not very reliable for the type of load presented by the linear induction motor operating at large airgaps. In fact induction motors operating at airgaps of 0.1 m to 0.3 m are not a practical proposition, or at least not compatible with the advantages of the much larger clearances possible with the electrodynamic levitation schemes.

The linear synchronous motor was at first ignored because of the requirement of synchronisation between the speed of operation and the frequency of its supply.

It was also assumed that the track structure – either the long stator or the rotor – would be expensive. The advent of superconductivity has changed the situation completely so that the synchronous motor now appears to be the key to propulsion at high speeds. The superconducting levitation magnets provide fields that are both intense and of large volume so that they can be coupled efficiently to fixed stator windings on the guideway, even at clearances of the order of 0.1 m to 0.3 m. The increase in the rotor current, i.e. the on-board superconducting magnets, opens the way to a reduction in the stator currents and so to a simpler and less costly stator structure than possible hitherto, such as an air cored stator. As to variable frequency inverters or converters, the recent advances in power transistor technology have also opened up the possibilities of light, compact, efficient, and reliable variable frequency sources, either alongside the track or on board.

Referring to Fig. 3.8(a), which shows the drag force on a magnet moving at constant height above a conducting track, the low speed drag peak occurs in the speed range 8 km h^{-1} to 16 km h^{-1} (5 mph to 10 mph), depending upon coil geometry and the thickness of the conducting track. At high speeds the drag falls off inversely as the speed v.

If the vehicle remains on wheels up to some lift off speed v_0 the magnetic drag on the vehicle is similar in shape to Fig. 3.8(a), provided that $v_0 > 80$ km h^{-1} and that the magnets remain at constant height for $v < v_0$. For a 50 tonne vehicle the estimated magnitude of the low speed drag peak is 8.9×10^4 N (20 000 lb) inclusive of the drag associated with the guidance magnets (approximately half of the lift forces). The total drag force on such a vehicle due to both aerodynamic and electromagnetic forces is shown in Fig. 3.11. This force is calculated on the assumption that the vehicle is entirely levitated. The aerodynamic drag is proportional to v^2

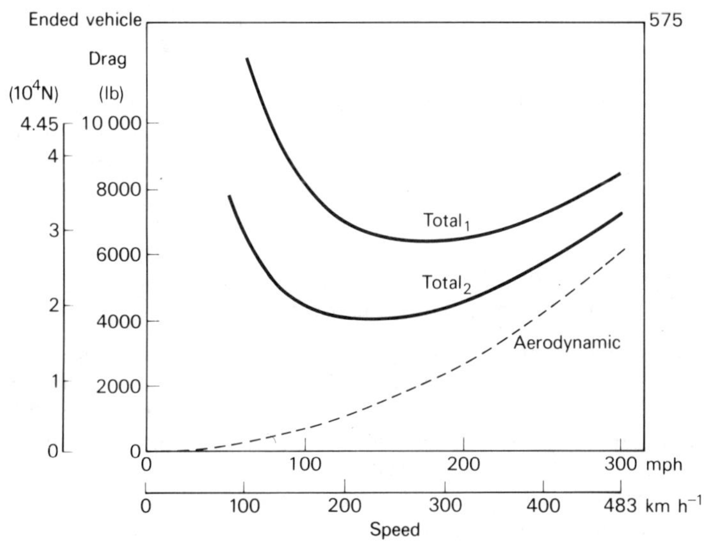

Fig. 3.11 Propulsion force requirements to overcome eddy current and aerodynamic drag as a function of speed.

and at 483 km h^{-1} (300 mph) is estimated[128] to be 2.7×10^4 N (6000 lb). For the electromagnetic forces two different cases of $F_L/F_D = 40$ and $F_L/F_D = 80$ at 483 km h^{-1} are considered. The drag power at 483 km h^{-1} (300 mph) is therefore 3.5 MW for airdrag and 1.5 MW for the electromagnetic drag if $F_L/F_D = 40$. At 80 km h^{-1} F_D is approximately 6.7×10^4 N (15 000 lb) for the same F_L/F_D at 483 km h^{-1}. Thus it would appear that the drag is appreciable over the entire speed range. There are therefore considerable problems in accelerating the vehicle with a thrust limited engine such as a linear induction motor. Obviously the propulsion force must exceed the drag force otherwise the vehicle will never achieve high speed operation and it might not be feasible to accelerate through the drag peak. It might however be more reasonable to look at the possibilities of reducing the drag peak.

Methods suggested for reducing or eliminating the low speed drag are as follows.

i) The magnets should not be operated in a persistent mode, thus allowing the current to be reduced during acceleration. The current is increased to full strength just before v_0 is reached. The magnet coils, however, will possess substantial inductance and it could take as much as 10 s to bring the current up to full value. Some additional cryogenic losses would also be introduced.

ii) The conductor should be left out of the critical acceleration section of the guideway, and in order to avoid an abrupt increase in the drag force, the conducting track should be brought up to wheel level gradually from below.

iii) The aluminium plate in the track may be tapered, although the drag peak cannot be avoided this way. However, the vehicle can pass through the drag barrier of a tapered section if it has sufficient velocity. Thus the vehicle could at first be accelerated to 80–110 km h^{-1} in a guideway without a metal plate. When it enters a guideway section with a tapered plate it will continue to accelerate. Finally it will reach the appropriate thickness for high speed operation and will be well beyond the drag peak. If the tapered section is limited to 15–30 m, the reduction in speed in passing through the drag peak is only 1.5–3 km h^{-1}.

iv) The vehicle may be fitted with wheels which maintain the magnets at a height above the guideway which is greater than the operating gap. Here all the electromagnetic forces including the drag peak will be substantially smaller. When the vehicle reaches a speed beyond the lift-off speed or the drag peak, wheels are retracted and the vehicle will then be supported magnetically. Although it adds the mechanical complexity of retractable wheels, this scheme has the advantage of being able to operate anywhere on the guideway, provided that provision is made for the wheels.

Calculations of braking from 483 km h^{-1} following loss of propulsion for the typical case of $F_L/F_D = 40$ at 483 km h^{-1} and $v_0 = 80$ km h^{-1} show that total acceleration takes about 3 minutes. Aerodynamic drag is assumed to be 2.7×10^4 N as previously. The peak decelerations, depending on the coil geometry and v_0, the lift speed, range from $0.17g$ to $0.3g$. Peak deceleration is reduced either by increasing v_0 or by increasing the length/width ratio of the coils. Increasing the conductor thickness in the guideway is equivalent to increasing v_0 since it is v_0/ω which is important. These figures of deceleration correspond roughly with the TACV (tracked air cushion vehicle) specifications of normal deceleration not exceeding $0.15g$ and emergency deceleration not exceeding $0.3g$.

As has been mentioned before, propulsion of a magnetically suspended vehicle requires considerable thrust. For example, it was indicated that if the lift speed v_0 is $80\,\text{km h}^{-1}$ and $F_L/F_D = 40$ at $483\,\text{km h}^{-1}$, the magnetic drag is $6.7 \times 10^4\,\text{N}$, and this drops to approximately $4.5 \times 10^4\,\text{N}$ (10 000 lb) for $v_0 = 120\,\text{km h}^{-1}$. To accelerate at $0.1g$ requires another $4.5 \times 10^4\,\text{N}$ of thrust. One of the largest linear induction motors was designed by Garrett Research Corporation. This was an 8000 HP (5970 kW) motor for the tracked air cushion vehicle. The design specification for this vehicle called for $6.7 \times 10^4\,\text{N}$ peak thrust for 1.5 minutes. The continuous cruise thrust specification for the Garrett motor was $4.5 \times 10^4\,\text{N}$. Whilst this would be adequate for a magnetically levitated vehicle to cruise at $463\,\text{km h}^{-1}$, clearly it is inadequate to overcome the drag peak. Another disadvantage of the LIM for suspended vehicle application is likely to be its weight. The unit above is estimated to weigh 12 tonne which is a substantial fraction of a 50 tonne vehicle.

3.3.4 Propulsion schemes using linear synchronous motors

It has been pointed out that the propulsion unit most compatible with large airgap operation of electrodynamic levitation systems appears to be a linear synchronous motor. In this system a travelling magnetic field produced by an active track interacts with large on-board superconducting magnets. Although in principle these magnets could be used for levitation and propulsion, schemes currently under construction in Japan and Canada (Figs 3.12 and 3.13 respectively) appear to have separate levitation and propulsion magnets.[40,43,131] The aspect ratio (width/length) of the coils for the suspension magnets should be small, whereas for the propulsion magnets it should be equal to or even greater than the one for suspension magnets. Since currents in the superconductors are large (3×10^5 to $5 \times 10^5\,\text{A}$)

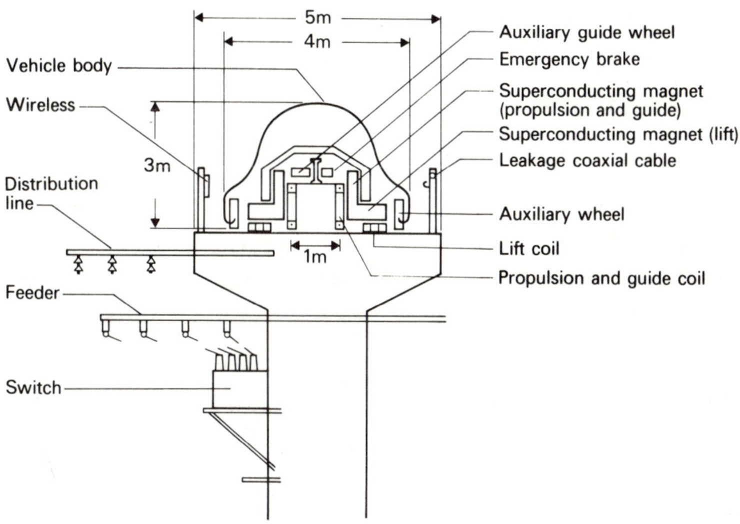

Fig. 3.12 Sectional view of Japanese National Railways second superconducting levitation vehicle.

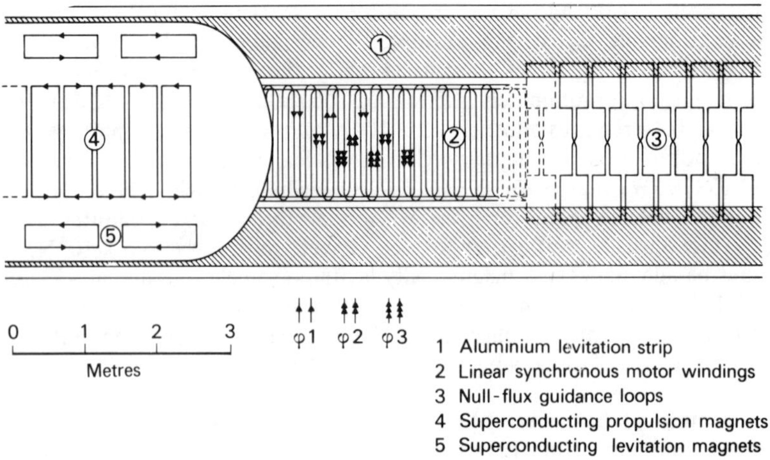

Fig. 3.13 A schematic plan view of the Canadian vehicle and guideway showing the superconducting levitation and guidance magnets and the linear synchronous motor windings with superconducting excitation magnets.

only about 1000 A to 1500 A are needed in the track windings. The higher the superconducting currents, the lower the track currents with correspondingly high efficiencies. Typical cruise efficiencies may be as high as 80% depending upon the length of the superconducting loops, the amount of current carried and the length of the energised track. Thrust in excess of 8.9×10^4 N (20 000 lb) is possible with a clearance of 0.1 to 0.3 m. The disadvantages are obviously the cost of the track and the complexity of providing an active track, but the advantages are that no heavy on-board motor is required, power collection of the order of megawatts at 483 km h^{-1} (300 mph) is avoided, the clearance is large, and control of many vehicles may be achieved. Additionally, the dual LSM proposed in the Canadian programme could provide active control of heave, pitch, roll and yaw motions of the vehicle.

3.4 Some basic configurations proposed or under construction

It is instructive to study the configurations and details of some of the electrodynamic (EDS) schemes under investigation. Currently only one scheme, that of the Japanese National Railways, is actively being pursued, but an examination of the various schemes tends to highlight many problems which remain unresolved. It also tends to indicate the direction in which further development must proceed.

The first proposal for repulsion levitation of vehicles by superconductors was published in 1966 by Powell and Danby[30] of the Brookhaven National Laboratory. As physicists, they were concerned with the design of magnets for confining, focusing and bending beams of high energy charged particles. Although superconducting magnets were then being used by physicists, the broader scientific community had neither the interest nor the money for application of superconducting magnets to transportation. The transportation community viewed the proposal more with amusement than with any genuine interest, and Powell and Danby developed the concept in their spare time. Their first scheme involved a superconducting guideway but they later refined what is known as the 'null flux' method which also appears to have been investigated in Germany. In the null flux method, a significant reduction in drag can be achieved by arranging the guideway loops so as to obtain an extremely high field gradient. This is accomplished by having two opposing guideway loops flanking a single coil on the vehicle or two opposing vehicle coils flanking a single guideway loop (see Fig. 3.14). A null flux system provides both a high lift-to-drag ratio and strong restoring forces or suspension stiffness. In fact the system is so stiff that the alignment guideway might become critical and the ride uncomfortable without secondary suspension. The secondary suspension might add considerable weight and create additional aerodynamic drag and power loss thus cancelling out much of the advantage.

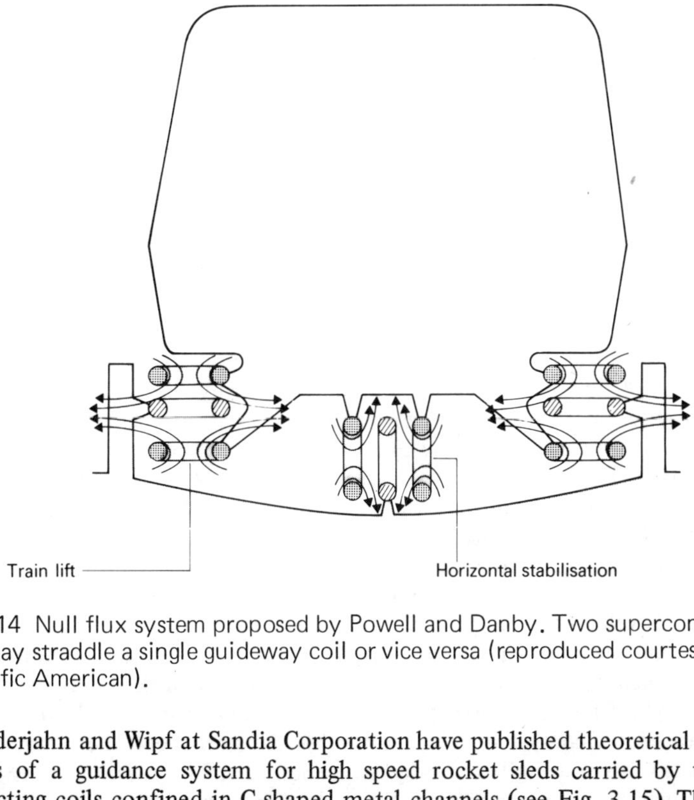

Fig. 3.14 Null flux system proposed by Powell and Danby. Two superconducting coils may straddle a single guideway coil or vice versa (reproduced courtesy of Scientific American).

Guderjahn and Wipf at Sandia Corporation have published theoretical and design studies of a guidance system for high speed rocket sleds carried by two superconducting coils confined in C-shaped metal channels (see Fig. 3.15). This was the first proposal for a continuous sheet guideway. The system would provide a very stiff suspension and high drag since the field is almost entirely perpendicular to the guideway sheets. Guderjahn and Wipf claimed later that the lift-to-drag ratio could be improved by adding a layer of steel to the upper portion of the guideway channel so as to derive a part of the lift by ferromagnetic attraction. A hybrid scheme of this kind has been adopted in the most recent variation developed by Powell and Danby, and Uranker[49] in 1974 elaborated a few other variations of the hybrid scheme. He claims advantages such as low drag losses, adjustable suspension

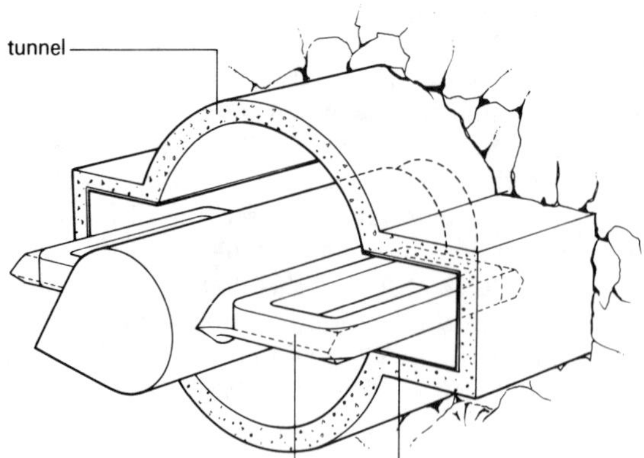

Fig. 3.15 Schematic of a rocket sled levitated using superconducting magnets.

stiffness, lower excitation currents, and hence lighter magnets and mechanical superstructure, etc. The examples of schemes suggested by Uranker are shown in Fig. 3.16(a) and the lift forces and drag losses as compared to the normal flux system in Fig. 3.16(b), for the three configurations underneath.

For the T-shaped track combining levitation by normal flux and guidance by null flux, it appears that the guidance system behaves like the basic null flux system but the specific losses of that part of the system causing lift are about half those of the usual null flux system. The absolute lift of this hybrid T-system corresponds to about one sixth the lift of a normal flux excitation magnet with the same excitation but a width equal to the separation between the magnets of the hybrid T-configuration. The hybrid T-configuration is interesting because separate optimisation of lift and guidance satisfies the requirement that the lift should be approximately twice the guidance force. The theoretical computations of the ferromagnetic hybrid system of Fig. 3.16(a) are also interesting, and are shown in Fig. 3.17.

The specific drag loss for constant load operations decreases at first, but later increases with velocity as in the basic null flux system. This is due to the rapidly increasing contribution from the ferromagnetic part of the system with decreasing distance between the magnets and the iron sheet. In spite of the lower specific drag loss of the hybrid ferromagnetic system the suspension stiffness is lower than that for the basic null flux system. The addition of the steel sheet, therefore, both reduces the stiffness and limits critically the region of stability. The suspension stiffness has values between the normal flux and null flux schemes. The distinct dependence on speed of the stiffness of the null flux system, however, is retained, as is shown in Fig. 3.17, and it is seen to vary from 0.1 $\mathrm{kN\,mm^{-1}}$ at 20 $\mathrm{m\,s^{-1}}$ to 0.8 $\mathrm{kN\,mm^{-1}}$ at 150 $\mathrm{m\,s^{-1}}$. A typical value quoted by Uranker for a 2 m long magnet travelling at 150 $\mathrm{m\,s^{-1}}$ is 2.1 $\mathrm{kN\,mm^{-1}}$.

There were, in the 1970s, two programmes in progress in Germany, one concerned with a superconducting levitation system and the other with an attraction DC electromagnetic system. The EDS system, which unfortunately has now been terminated, was being investigated at Erlangen on a 280 m diameter track (see

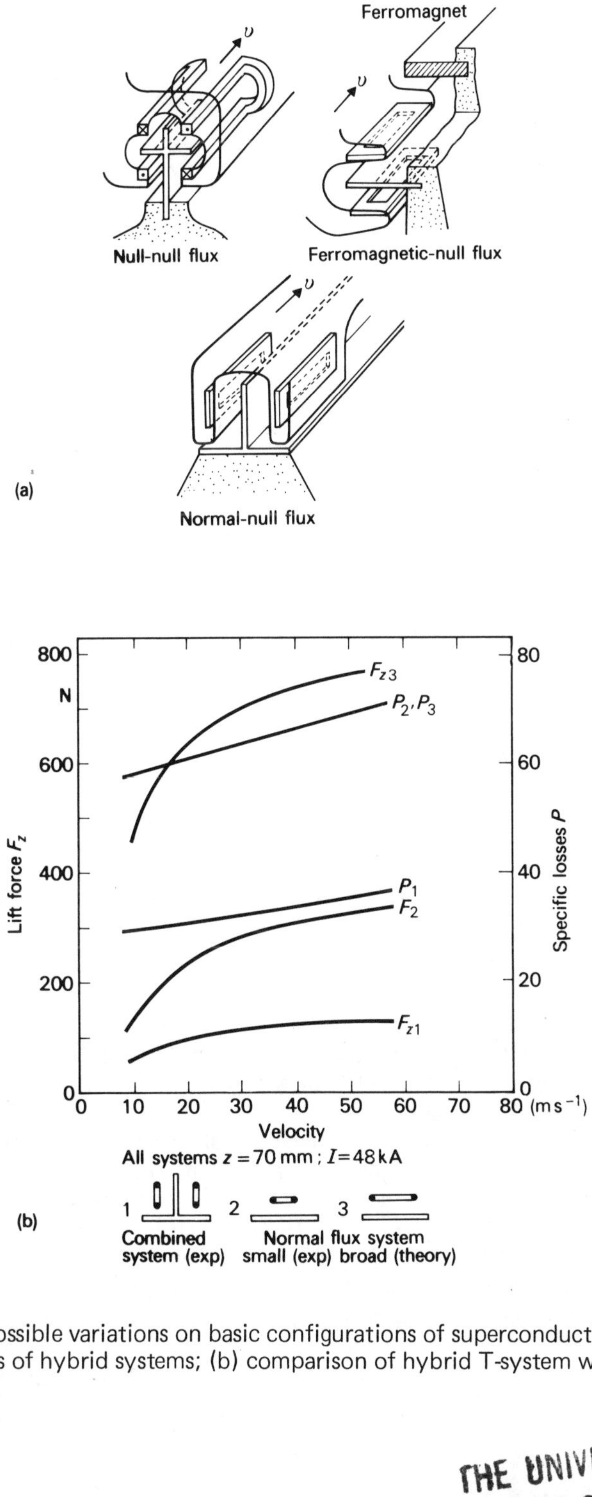

Fig. 3.16 Possible variations on basic configurations of superconducting systems: (a) examples of hybrid systems; (b) comparison of hybrid T-system with normal flux system.

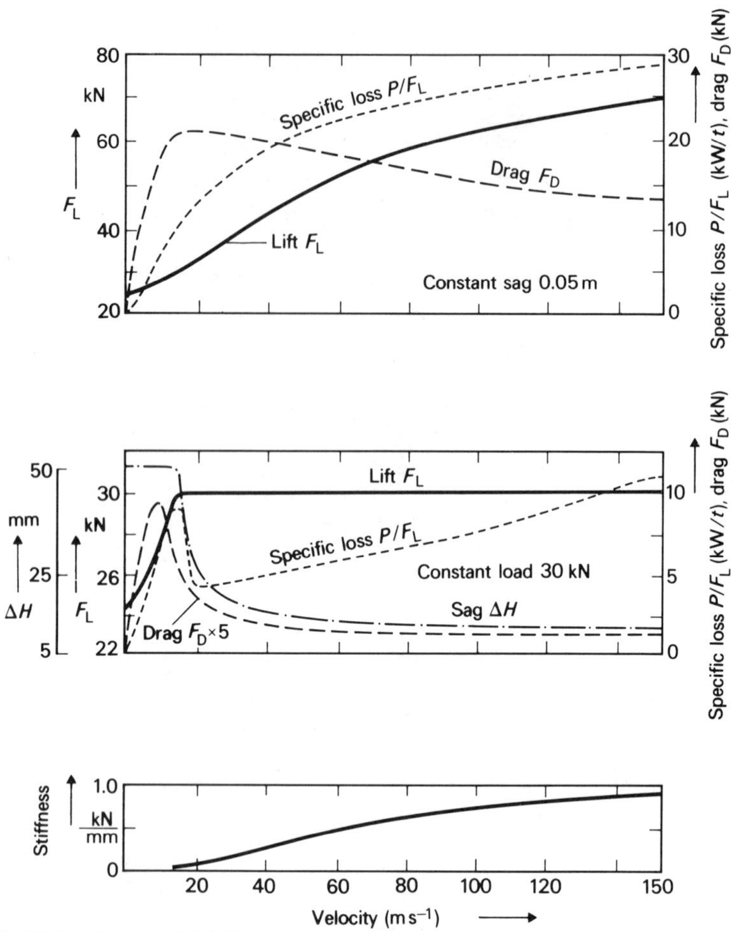

Fig. 3.17 Variation of (a) lift force drag, and lift power to lift force ratio, for constant gap; (b) lift force, drag force, lift power to lift force ratio with airgap variation; (c) stiffness of suspension with speed.

Fig. 1.6) of double-C cross section banked at an angle of 45° so that there were no lateral forces at 140 km h^{-1}. The test vehicle (see Fig. 3.18) was 10 m long and weighed 12 tonnes. It was intended to operate at 200 km h^{-1} at an airgap of 10 cm, propelled by a linear induction motor with a rated thrust of 22 kN. The superconductive magnets are carried on two magnet frames, one at the front and one at the rear of the vehicle. These magnet carriers are moveable and can be displaced upwards and downwards by hydraulic systems. Each of these frames contains four superconducting magnets, two for lifting and two for guiding the vehicle. The vehicle runs on wheels and when it has acquired sufficient speed the magnet carriers are lowered towards the track when levitation commences. The eight magnets are disc shaped with dimensions length, 1.3 m, width, 0.4 m and height, 0.25 m. Each magnet weighs 350 kg and is capable of generating maximum lift and guidance forces of 100 kN at 10 cm gap. The lift magnets have additional normal windings

(a)

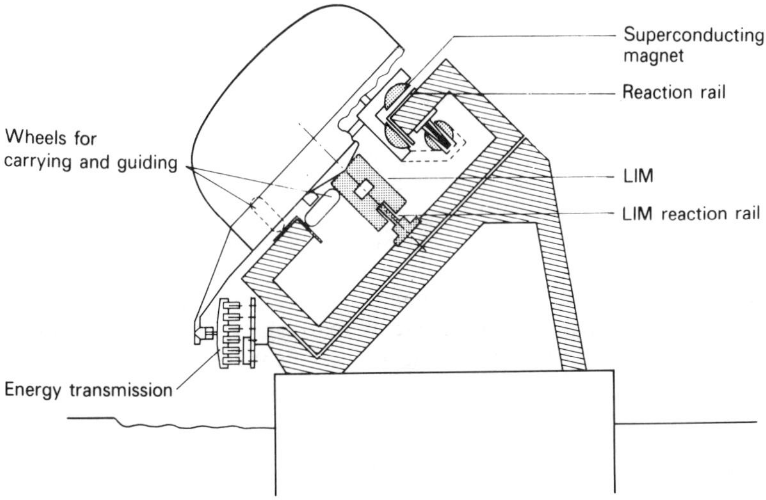

(b)

Fig. 3.18 AEG, BBC and Siemens superconducting magnet vehicle project: (a) vehicle and track at Erlangen (photograph courtesy of Siemens Ltd); (b) cross section of the track and vehicle.

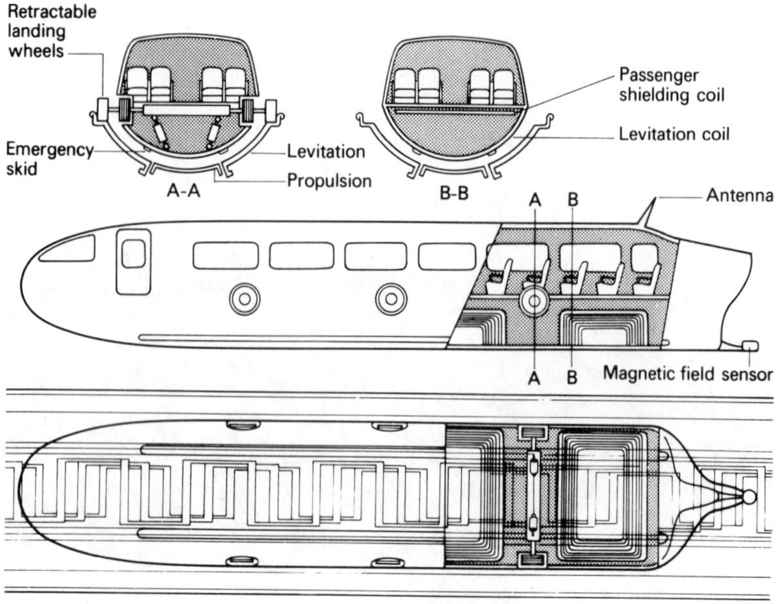

Fig. 3.19 Sectional views of the Magneplane vehicle developed by Kolm and Thornton at the MIT (reproduced courtesy of Scientific American).

which are used as control windings to provide active damping and ride control. The work at Erlangen now appears to be exclusively devoted to investigation of the air-cored linear synchronous motor.[131] The test carrier is equipped with two superconducting magnets. These act as excitation magnets for the aircored winding on the track.

Work on the 'Magneplane' scheme was initiated in 1969 at the MIT. The essential outlines of the magneplane vehicle developed by Kolm and Thornton[132,133] are shown in Fig. 3.19. It was supported by saddle-shaped superconducting coils distributed essentially over its entire underside. It was intended to levitate about 25 cm above the guideway when travelling at between 320 and 480 km h^{-1} (200 and 300 mph). The guideway was a semicircular aluminium trough which would allow the vehicle to assume a coordinated bank angle on curves. Self-banking was considered essential because with a maximum angle of 45° banking, the steepest performed by an aircraft, the tightest radius that can be negotiated at 483 km h^{-1} (300 mph) is 1500 m (one mile). The assumption here was that if the banking angle were fixed by the guideway, even slight variations in speed would lead to unacceptable side forces on the passengers. The propulsion system adopted for the magneplane was the linear synchronous motor with an active guideway. The form of the stator winding, consisting of meandering conductors between the levitation conductor sheets, is shown in Figs 3.19 and 3.20. In this guideway design most of the aluminium served both structural and electrical purposes. It was claimed, therefore, that the cost of the guideway was comparable to that of any other projected high speed system and not in any way dominated by the fact that the guideway and the propulsion system are integral.

Synchronisation of the linear synchronous motor was to be achieved by

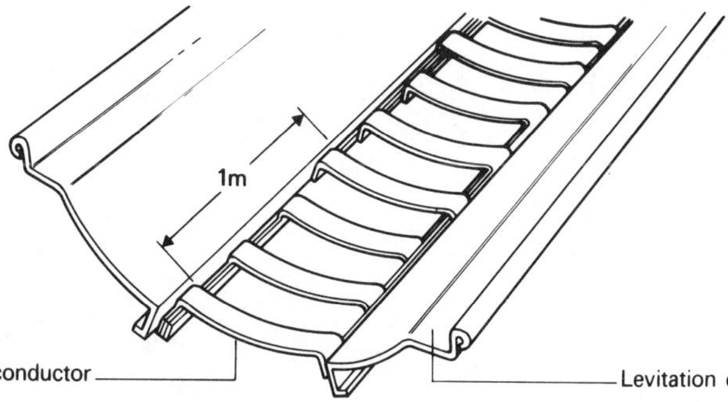

Fig. 3.20 Magneplane guideway with synchronous linear motor conductors on the guideway. The passive aluminium troughs are intended to allow the vehicle to roll into a coordinated bank angle in curves. The vehicle carries the superconducting excitation magnets of the LSM (reproduced courtesy of Scientific American).

transmitting position information from the vehicle to the wayside power control units to be positioned at say 10 mile intervals on the full-scale system. The units would contain transformers and computers that actuate the transformers and solid state switches positioned at one mile intervals along the guideway. The phase, frequency and amplitude of the polyphase propulsion currents would be adjusted automatically to maintain synchronous acceleration, deceleration or constant speed. It was thus claimed that the system substituted information transmission for power transmission and that there were also some further advantages. The first was that control of vehicles by wayside units connected to a central computer allows safe operation at the levels of traffic density and vehicle headways necessary for economic operation of these systems. The second advantage was that propulsion system control can be used actively to damp out the inherent slow oscillations of the order of 1 Hz. Schemes have been proposed[134] elsewhere also for control of heave, roll, pitch and yaw.

Studies and some experimental work are in progress at the Canadian Institute of Guided Ground Transport (CIGGT), to investigate electrodynamic levitation systems and linear synchronous motor propulsion.[135] The experimental work consists mainly of a wheel 7.6 m (25 ft) in diameter rotating about a vertical axis at peripheral speeds up to 101 km h^{-1} (63 mph) driven by a 1750 rpm, 120 kW variable speed DC motor (see Fig. 3.21) through a 25 : 1 reduction gear box. Guideway components, i.e. conductors, can be attached to the vertical rim while vehicle-borne components, i.e. full-scale superconducting magnets, are mounted in a stationary harness, and their interactions are measured by a six component balance with positional adjustment torques.

The Canadian investigation is aimed at producing by the 1990s electrodynamically levitated vehicles with linear synchronous motor propulsion and guidance for 480 km h^{-1} inter-city transit along the Toronto-Ottawa-Montreal corridor. The vehicles are intended to carry 100 passengers and to weigh 30 tonnes, levitated with 15 cm clearance by eight vehicle-borne superconducting magnets. The lift magnets are 1.0 m long and 0.3 m wide with 3.85×10^5 ampere turns interacting with eddy currents induced in 80 cm wide aluminium strips on a flat topped guideway.

Fig. 3.21 The Canadian Universities (CIGGT) test wheel for superconducting levitation systems: photograph of the wheel showing the LSM windings and an adjacent null flux loop (courtesy of the Canadian Institute of Guided Ground Transport).

The aluminium strip thickness is graded from 1 cm at high speed to 3 cm in low speed sections to maintain total drag (magnetic and aerodynamic) almost speed independent whilst minimising the combined costs of energy requirements and aluminium amortisation. For the synchronous propulsion system the vehicle is supposed to carry fifty superconducting magnets, each 0.4 m long and 1.5 m wide. The guideway is to have a split three-phase winding energised by variable frequency current source inverters in 5 km sections to give 72% efficiency and 0.73 power factor. The thrust force is controlled by sensing the phase angle between supply current and induced voltage in the guideway windings, thus minimising the inverter rating and introducing the possibility of dynamic control of the vehicle.

A flat topped guideway is proposed to minimise ice and snow accumulation and a new technique is proposed for obtaining lateral guidance by using interaction of the 50 propulsion magnets with the edges of the levitation strips and with null flux loops overlaying the LSM windings. The stiffness expected to be obtained is $10^6\,N\,m^{-1}$. The outline of the proposed vehicles is shown in Fig. 3.22.

The last but not the least of the developments in electrodynamic systems is that of Japanese National Railways (JNR) because it appears at the time of writing that this is the only project being actively pursued. Two vehicles have been built[41] and the first one was tested in 1972. The outline of this first vehicle is shown in Fig. 3.23. It was 7 m long, 2.5 m wide, 2.2 m high and weighed 3.5 tonne. It was propelled by a linear induction motor which had its secondary member on the vehicle and primary member on the track. The track was only 480 m long and hence the

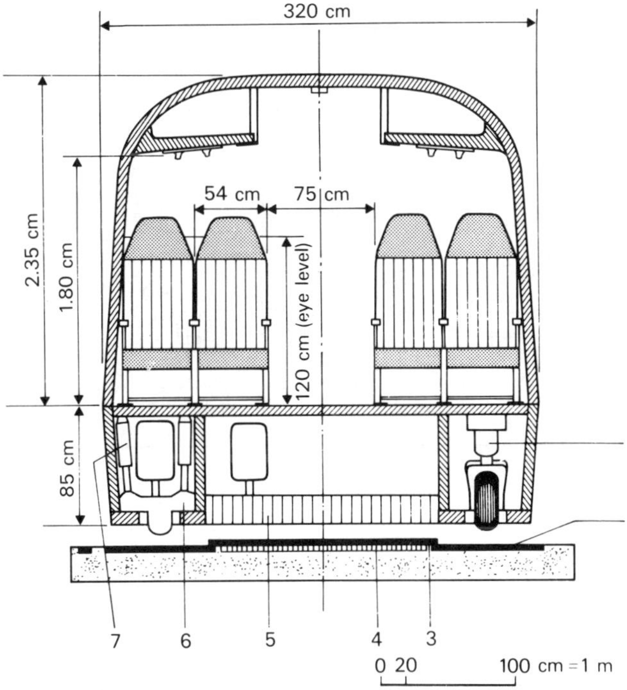

1 Retractable wheel for low speed suspension
2 Aluminum guideway surface for levitation
3 Guidance coils
4 Linear synchronous motor windings
5 Propulsion superconducting magnet
6 Levitation superconducting magnet
7 Secondary suspension

Fig. 3.22 A schematic section of the proposed Canadian superconducting levitation vehicle and guideway.

maximum speed attained was limited to 65 km h^{-1}. The clearance at this speed was 6 cm and lateral guidance was provided by wheels operating against side rails.

The second vehicle built by JNR and being tested on a much longer track of 20 km was finished in 1978 and it is therefore too early to obtain any published results for its operation. However, it is known that this second vehicle was to be 10 m long, 3.8 m wide and 2.7 m high with a weight of 10 tonne. It is illustrated in Fig. 3.24. The lifting superconducting magnets have 1.5×10^5 ampere turns each and are installed horizontally on the vehicle. The secondary members are aluminium coils on the guideway (the configuration adopted is in the form of a ladder track) and the lifting magnet system acts in the normal flux mode. Another set of superconducting magnets, each with 4×10^5 ampere turns, is installed vertically on the vehicle and this set of magnets produces thrust and lateral guidance forces acting on vertical coils attached to the guideway. The guidance magnets act in the null flux mode. The vertical coils are also the primary of the linear synchronous motor

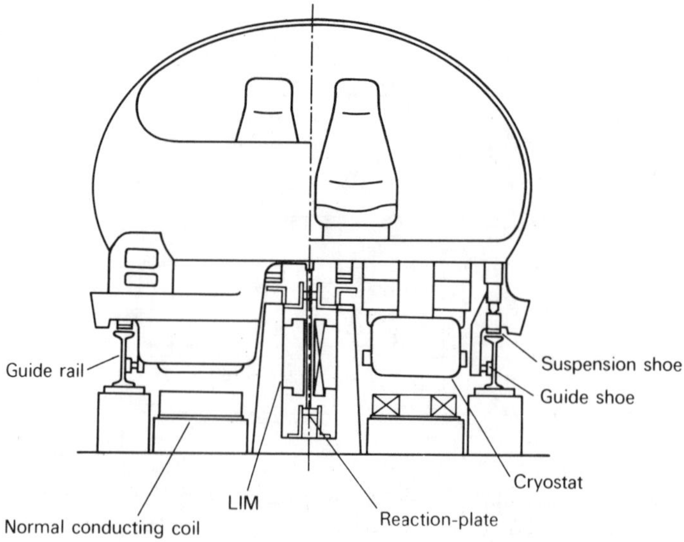

Fig. 3.23 Sectional view of Japanese National Railways first superconducting levitation vehicle ML-100.

Fig. 3.24 Photograph of Japanese National Railways second superconducting levitation vehicle undergoing trials on its track near Kyushu (courtesy of Japanese National Railways).

and are energised by a cycloconverter in the frequency range 0–33.3 Hz. Synchronism of the LSM is maintained by optical means.

3.5 Summary

Various experimental and a very large number of theoretical studies on the superconducting levitation systems have been carried out since the proposal in 1966 by Powell and Danby. The questions of dynamics and ride quality of vehicles are dealt with in Chapter 5 of this book but it appears from a study of the available literature that the early expectations of investigators, i.e. that superconducting magnet levitation systems would result in vehicles which do not require any active control, will not be realised. The basic problem here seems to be a highly underdamped suspension which results in oscillations of the order of 1 Hz. Of the active control schemes proposed the ones employing linear synchronous motors in a dual role seem to be in favour although none appear to have been tested. The problems of propulsion also seem to be quite considerable. In particular, on-board propulsion units (primaries) with sufficient thrust to overcome the low speed drag peak on the one hand appear to impose a severe weight penalty, but on the other an even more formidable problem common to all levitated high speed vehicles appears to be that of power collection at speed. The use of linear synchronous motors with active guideway requires a fairly complex control of sectionalising, and although this method is technically feasible its economic feasibility is doubtful for a revenue paying transportation system. Some doubts also remain about the reliability of superconducting magnets and cryogenic systems operating in transport applications but again these are not technically impossible to solve. On the whole, it seems that the technological problems still unsolved are of considerable magnitude. If they can soon be tackled satisfactorily, it is possible that around the turn of the century, superconducting magnetic levitation will be a strong contender for high speed transport systems.

4

Levitation using eddy currents induced by mains frequency excitation

4.1 Introduction

Conducting materials in the solid and liquid state can be levitated above AC coil systems. The levitation of solid plates or rings has been used as a laboratory or lecture demonstration.[53] The levitation of molten metal[52] for zone refining dates back to the early 1950s, when there was a considerable activity in the general area of induction machines, and is now a standard laboratory procedure. The notion that this form of levitation could be applied to passenger carrying vehicles appears to have originated towards the end of the 1960s or early 1970s[54,56,136] as a result of several years of study related to the use of linear induction motors as propulsion units for high speed vehicles.[137,138] Although the problem mentioned in Chapter 3, i.e. the difficulty of power collection at speeds in excess of 483 $\mathrm{km\,h^{-1}}$ (300 mph), is just as relevant in this context, the idea that linear induction motors could be used both to propel passenger carrying vehicles and to levitate them gained considerable attention and popularity in the mid-1970s. The idea became known as the 'magnetic river'.

A single sided linear induction motor can be designed to produce large levitation forces in addition to its designed translational or tractive force. As the speed of the linear motor increases with respect to a composite reaction plate consisting of a conducting material with a permeable backing material such as steel, the force of repulsion becomes a force of attraction in a normal design. The concept of magnetic river revolves round the objective of the design of a single sided linear motor which could combine the functions of propulsion, suspension, and guidance. Further in such a motor the force of repulsion would remain constant up to running speed with little or no additional input.

4.2 AC induction levitators

4.2.1 Some unstable configurations

The study of AC levitation is closely linked with that of induction machines. For example, the lifting force on a conducting sheet over the surface of a single sided linear motor can be calculated[139] provided that the equations for the flux density b and current density j at all points can be found. However, in most practical levitation systems the equations either cannot be formulated or are so formidable that solutions are virtually impracticable. A qualitative appreciation based mostly on experimental work is therefore highly desirable.

A jumping ring experiment illustrates adequately the forces of repulsion between a conducting loop threading an open iron core produced by AC in a single coil on the same core. The geometry lends itself to fairly accurate calculation of the force if the leakage flux is assumed to decay exponentially[140] along the length of the core. The experiment was originally considered to be no more than a lecture or laboratory demonstration. Recently, however, an application was proposed for this in an aluminium smelting process for stirring the molten metal.[141]

The configuration of this experiment is shown in Fig. 4.1. If the core is long enough, the coil is energised and the ring takes up a steady state vertical position.

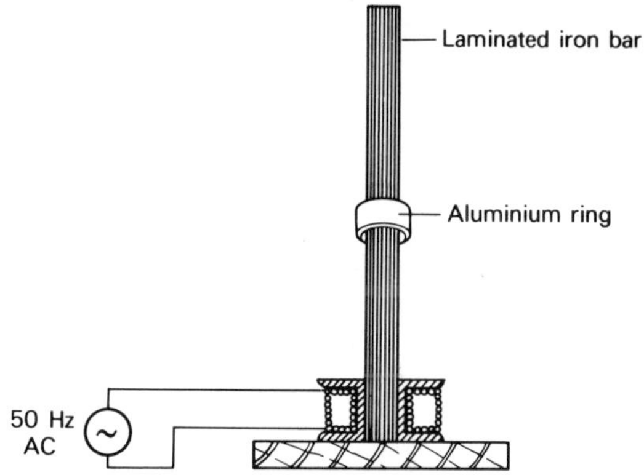

Fig. 4.1 The jumping ring experiment.

As shown in Fig. 4.1 it will be found that the ring is always in contact with the core somewhere along its perimeter. The ring is therefore not levitated stably, in that it is physically restrained by the core. The link between the study of conventional motors and their driving forces on the one hand, and the study of levitation using AC and lift forces on the other is provided to some extent by the shaded pole motors. In such motors the phase changes produced by the shading ring are capable of producing force on the ring itself. The transition from a conventional three-phase induction motor to a circular plate levitator through a shaded pole motor and a ring levitator is shown[53] in Fig. 4.2. Starting with a conventional three-phase motor in Fig. 4.2(a), in which the phase progression of the airgap flux along the perimeter of the rotor is dictated by the voltage to which the stator windings are connected, the first step is to replace the polyphase winding by a single-phase winding with shading ring(s) on the pole(s) (see Fig. 4.2(b)). Next the rotor conductors are replaced by a continuous cylinder (see Fig. 4.2(c)) and the machine unrolled into a linear machine (see Fig. 4.2(d)). It is then rerolled about the axis AB (see Fig. 4.2(d) and (e)) with the stator winding on the inside; the thrust is now such as to push the conducting cylinder out of the stator. The motor is then turned through 90° so that the force on the cylinder is now vertically upwards. The length of the machine is reduced so that the number of windings and shaded rings is reduced to the minimum (see Fig. 4.2(f)). The next step, shown in Fig. 4.2(g), is important and involves using the shaded ring to produce a force on itself. The jumping ring geometry is one in which the outer iron cylinder is removed. The action is still that of a travelling field moving up the annular slot.

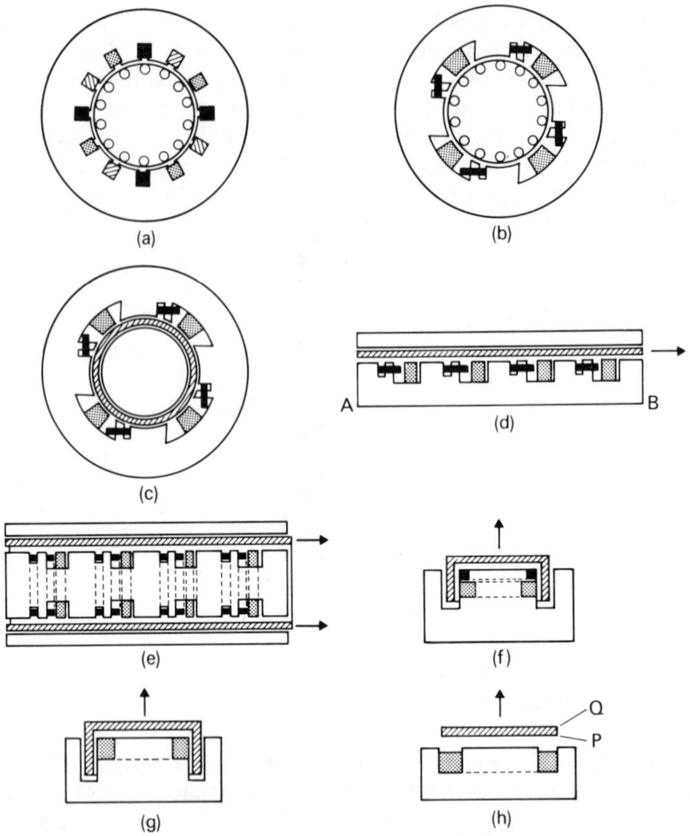

Fig. 4.2 Evolution of an eddy current repulsion system from a 3-phase squirrel cage motor: (a) conventional rotary machine; (b) shaded pole motor with a squirrel cage rotor; (c) shaded pole motor with a drag cup rotor; (d) linear shaded pole motor with a sheet rotor; (e) tubular shaded pole motor with axial flux; (f) single coil tubular shaded pole motor; (g) elementary levitator; (h) circular plate levitator (reproduced courtesy of the IEE).

The flux fringes above the slot opening, and thus the action does not stop abruptly at the opening: phase changes of current and flux also occur in the parts of the cup outside the slot. Currents circulate around the rim of the cup and ideally there are no currents in any other axis. The cylindrical rim of the cup is now removed leaving a circular conducting ring, the hollow part of which may now be filled (see Fig. 4.2(h)), thus completing the transition to a circular plate levitator. The currents which circulate around the lower part of the periphery in the position P are in phase advance over those circulating around the upper part in position Q, and the lifting force may be attributed to a travelling field pattern moving upwards and operating on the plate in the manner of an induction motor. It is relatively simple to show that the system is unstable. As soon as the plate is displaced slightly from the central axis (see Fig. 4.3), the forces originating from the pole shading

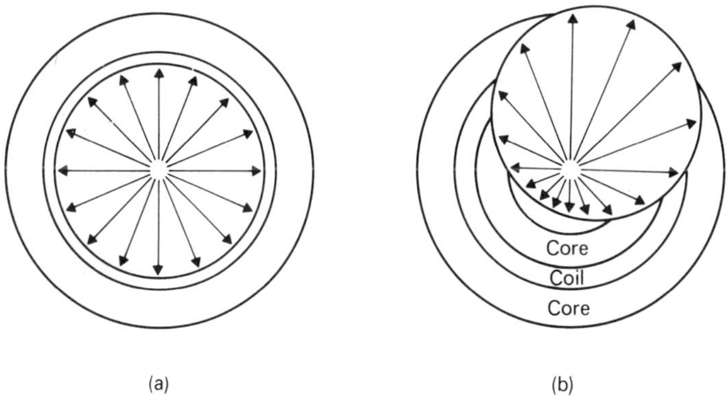

Fig. 4.3 Horizontal forces on a circular levitated plate: (a) plate concentric with the coil; (b) plate displaced off centre (reproduced courtesy of the IEE).

action will operate first on the overhanging side and then on the other, and the plate will be thrown off the coil-face.

4.2.2 Some stable suspensions

Circular plates

The reason for the disc in the plate levitator of Fig. 4.3 being thrown off may also be attributed to the presence of the plate producing radially outward travelling fields. Stability could therefore be achieved in principle by a number of concentric circular coils, each connected to a voltage of different phase and thus producing a travelling field moving radially inwards. In practice two such coils have been found sufficient to produce stability. If the core is made up of radial laminations, the tapered form being achieved by staggering them radially, the plan view of the core would be as shown in Fig. 4.4(a). More conveniently it could be as shown in Fig. 4.4(b) but made up into blocks. The cross-sectional view of the arrangement is shown in Fig. 4.4(c). Stable suspension can be obtained for a range of plate diameters between the outer diameter of the inner exciting coil and the inner diameter of the outer coil. The conditions for stability vary widely with the diameters, even within this range. Plates whose diameters are less than the outer coil diameter can be suspended stably with the centre of the plate displaced away from the centre of the coil. The plate is then also capable of being spun as if it were inside an invisible tube of a larger diameter than that of the plate itself. The principal design dimensions for the circular plate levitator described in reference 53 are height 14 cm (5.5 in) and overall diameter 20.3 cm (8 in), with lamination of width 1.3 cm (0.5 in). The number of turns on the inner coil is 480 and the number of turns on the outer coil is 440; the coils are made of 15 SG enamelled copper wire.

An investigation of the stability conditions for a disc of given size was carried out by controlling separately the currents fed to the two coils; the method of presentation of the results shown in Fig. 4.5 is due to Cashmore.[142] Each value of the inner coil current, which is held fixed for one set of observations, will produce a diagram such as Fig. 4.5. The following five different operational conditions are represented on this diagram:

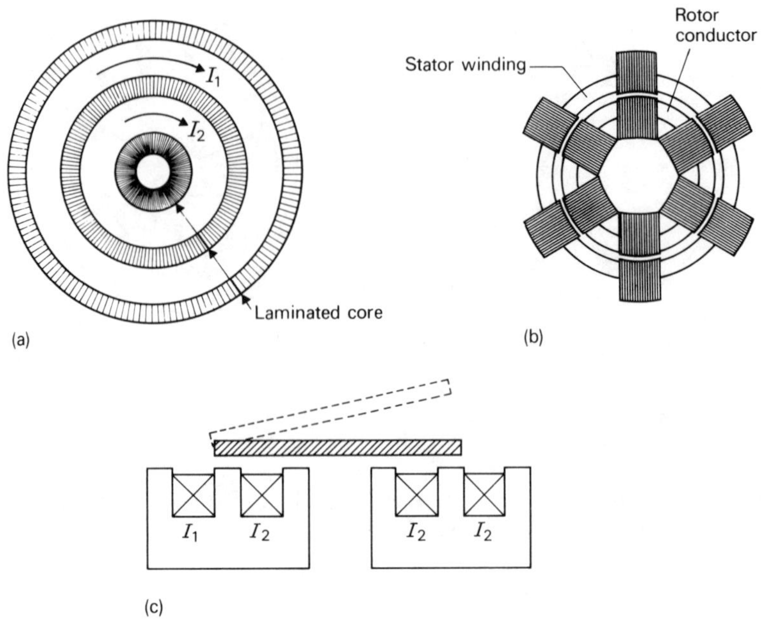

Fig. 4.4 Circular plate levitator: (a) laminated core and coils; (b) radial arm construction of the core; (c) relative positions of coils and plate (reproduced courtesy of the IEE).

(i) stable and level suspension;
(ii) continuous lateral oscillation of the disc resembling the oscillation of a steel ball in a trough;
(iii) oscillation of the disc with increasing amplitude until the disc is thrown out;
(iv) stable suspension with the plate tilted as shown dotted in Fig. 4.4(c);
(v) completely unstable with the plate thrown off as soon as it is released.

The transitions between the stable region (i) and the other regions (ii) to (v) are quite sharp, but other transitions such as those between conditions (ii) and (iv) are gradual with tilt and oscillation occurring simultaneously for quite a wide range. The dotted lines on these curves indicate the conditions when the tilt is large enough for the plate to touch the core. In exceptional cases the plate oscillates in the vertical direction, but the oscillation soon becomes lateral. This would appear to be analogous to the jumping ring experiment of Fig. 4.1. The ring can be made to oscillate[140] when the coil is connected in series with a capacitance of a value larger than that required to resonate the circuit with the ring midway between the coil and the end of the core. On some rare occasions rings made of high conductivity aluminium have continued to oscillate even when the capacitance has been switched out of circuit or at a particular setting of the variac supplying the coil.

For any particular value of the inner coil current the height of the plate above the core does not vary a great deal when the outer coil current is in the range where the levitated plate is stable. The formula given in reference 53, based on the assumption that for a given height the upward force on the plate is proportional to the

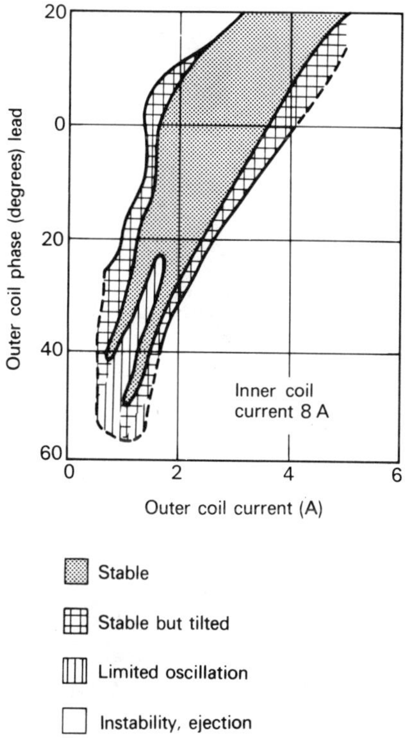

Fig. 4.5 Behaviour graph of levitator for inner coil current held constant at 10 A (reproduced courtesy of the IEE).

square of the primary current, states that lift at height h is proportional to $1/(h + \delta)^4$, where δ is a constant necessary to obtain the correct value of the lift at $h = 0$. The inner coil may be regarded as the lifting coil and the outer coil as the stabilising coil. Most of the losses in the plate, therefore, are supplied through the inner coil. The power input depends on various factors such as the disc thickness, the slot dimensions etc., but for a fixed configuration, such as the one for which principal dimensions are given earlier in this section, the power dissipated in the disc is a linear function of the height above the stator core. Again in reference 53, the formula for the power loss is given as $P_D = k_1 + k_2h$; the constants k_1 and k_2 can be measured in terms of the mass of material to be lifted. For copper these are quoted as $k_1 = 27$ and $k_2 = 36$ with h measured in centimetres and P_D in watts kg^{-1}; for aluminium they are quoted as $k_1 = 52$ and $k_2 = 32$ for the structure referred to. The total power input is obtained by adding the stator I^2R loss and the core loss to P_D. The I^2R loss increases as the fourth power of the height of levitation of the plate h. In the model described, I^2R loss is equal to the power dissipated in the plate at $h = 0.5$ cm.

The experimental plots such as those of Fig. 4.5 show a general pattern in relation to the behaviour of the plate just beyond the boundaries of the totally stable region. This includes the amount of tolerable displacement into the unstable

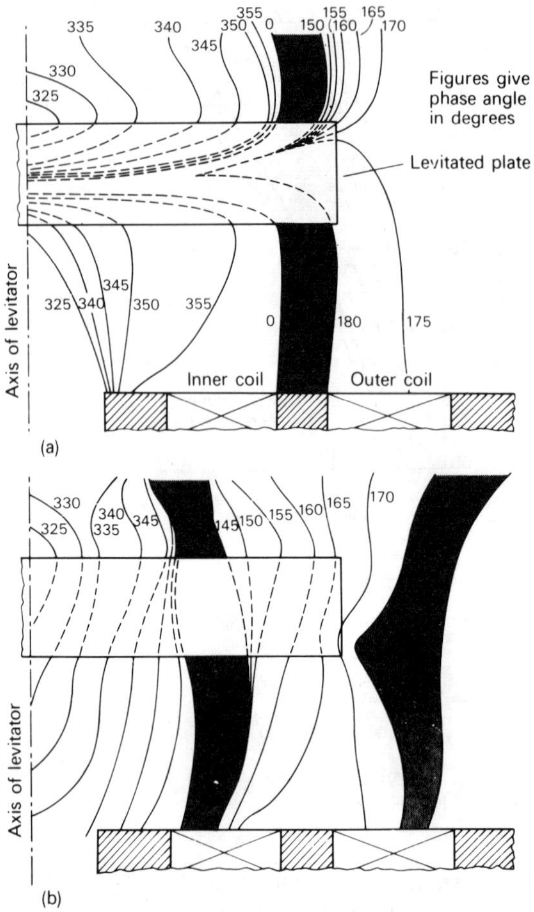

Fig. 4.6 Constant phase plots for circular levitator: (a) refers to B_x; (b) refers to B_y (reproduced courtesy of the IEE).

regions. Flux density plots in the space above the levitator when the plate is levitated, particularly the phase distribution, enable an appreciation of how the forces on the disc are produced and even how these forces are distributed throughout the volume of the conductor. Lines of constant phase indicate at different points the travelling fields which are responsible for the forces. Care should be taken not to confuse lines of constant phase with flux lines because they are often in the same direction. Figs 4.6(a) and (b) show constant phase contours for the components B_x (horizontal) and B_y respectively. Because the disc levitator is a circularly symmetric configuration, a cross section through the centre in a vertical plane depicting a two-dimensional plot is adequate to present all the necessary information. The notation adopted in these phase contour maps is that lines denoted by a lower number of degrees represents a flux lagging in phase behind one represented by lines with higher numbers so that the field can be said to travel towards the lines with the lower numbers. The phase contours are therefore analogous to height contours on a geographical map; movement towards a lagging phase line represents going downhill.

In addition to travelling fields there may exist rotational fields which may be detected by observing that the components B_x and B_y at a particular point are of different phase. This happens where a constant phase line from one set, bearing one number, crosses another one from the other set, bearing a different number. Hesmondhalgh[143] has shown that very thin conducting cylinders, so small as not to disturb the field pattern, mounted in bearings at the point in question will, in fact, rotate. Graphs of lines of constant phase for the components B_x and B_y are drawn from practical measurements and such graphs can then be used in conjunction with others depicting the magnitudes of the flux components to enable a force diagram to be constructed. The dotted lines of Figs 4.6 (a) and (b) show the probable combination of the contours through the conducting material. Figs 4.7(a) and (b) show the corresponding experimentally obtained[53] plots of flux magnitudes and these bear some resemblance to the flux plots of sinusoidal current distribution on the surface of a semi-infinite iron plate.[139] Combining the parts of Figs 4.6 and 4.7 produces a force distribution diagram such as the one in Fig. 4.8. The arrows on this diagram indicate the magnitudes and directions of the forces in the regions in which they are located. The plot suggests that the centre of the disc could be removed without affecting the force distribution, in particular the lift, a great deal. The inside radius of the annular ring is seen to be smaller than the inside radius of the inner coil.

Rectangular plates

The linear equivalent of the circular plate levitator, i.e. the one used to levitate rectangular plates, is an extension of the ideas discussed so far. Fig. 4.9 shows the evolution of the rectangular plate levitator from the circular disc levitator. The behaviour of the arrangement of Fig. 4.9(a) is very little different from that of a truly circular one. In Fig. 4.9(b) the number of blocks is reduced to four and a square plate of the appropriate size can be supported with this arrangement. There are only two energising coils in this arrangement so that the same currents flow through the slots of the four blocks. If the currents in each slot were returned beneath the individual blocks in the manner of a Gramme ring winding (see Fig. 4.9(c)), currents in the eight slots could be controlled independently, thus making the stability and attitude of the plate in each of the four directions x–x' and y–y' also independent.

The last step of this evolution is that of Fig. 4.9(d), in which $I_{y1} = I_{y2} = I_{y4} = 0$ and the blocks carrying currents along the x-direction have been elongated. If the system is found to be stable for $I_{x1} = I_{x2} = I_{x3} = I_{x4}$ with the directions of currents shown, the arrangement of Fig. 4.9(e) is identical but with the advantage that the two blocks are now self-contained and can be moved closer or further apart to accommodate plates of different widths. Just as there is no resistance to the motion of a spinning disc, there is no resistance to the motion of the plate along the x–x' direction and the movement corresponds to the rotation of discs about the vertical axis in the circular structure which is unresisted by the electromagnetic system, assuming there are no irregularities of construction.

Spheres and cylinders

Stable levitation of bodies with spherical symmetry is generally easier to obtain

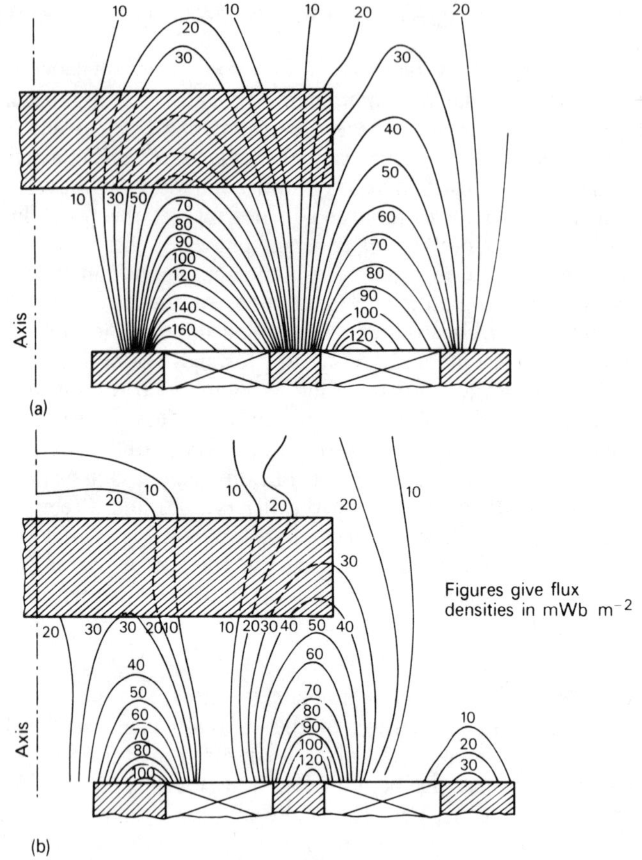

Fig. 4.7 Constant magnitude plots for a circular plate levitator: (a) refers to $\boldsymbol{B_x}$; (b) refers to $\boldsymbol{B_y}$ (reproduced courtesy of the IEE).

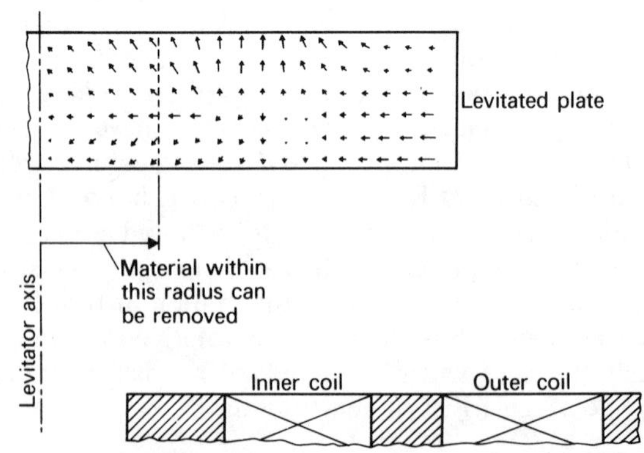

Fig. 4.8 Force diagram for plate levitator (reproduced courtesy of the IEE).

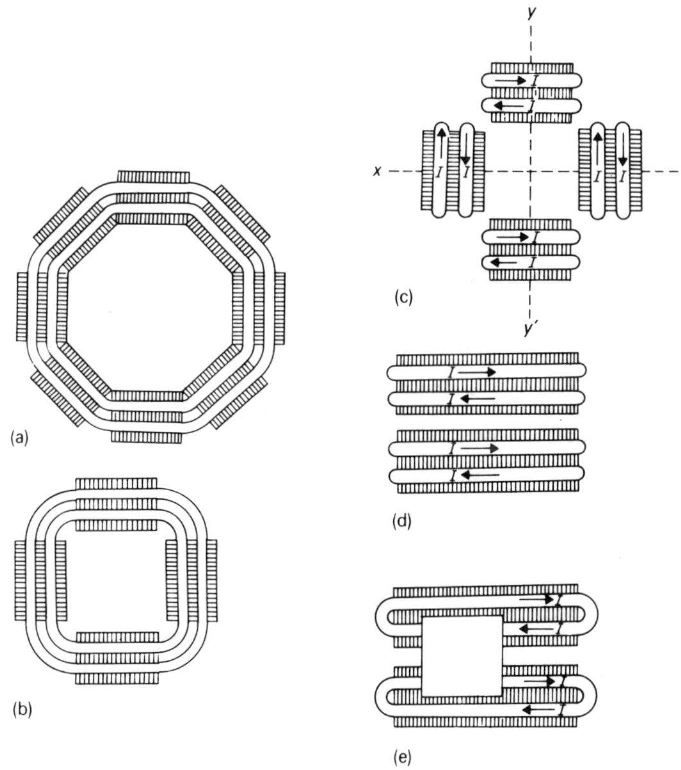

Fig. 4.9 Practical arrangements for disc levitators: (a) and (b) with two concentric coils; (c) with four coils on each axis; (d) blocks elongated in the x–x' direction for levitation of rectangular plates; (e) levitation scheme for rectangular plates (reproduced courtesy of the IEE).

than it is with flat plates. The construction of the levitator at power frequencies is basically the same as in the case of discs. It consists of two concentric coils in an iron structure and the inner coil can often be shorted out, thereby acting as a shorted ring instead of being fed from an external power source. Fig. 4.10 shows[53] the construction of a levitator in which the hollow aluminium sphere of wall thickness 0.75 mm ($\frac{1}{32}$ in) is levitated by a single coil which encloses a thick copper cylinder as a shading ring with a split iron cylinder separating the two. If thicker material, for example 4.5 mm ($\frac{3}{16}$ in) wall thickness, is used for the sphere, it may become oscillatory about a horizontal axis, finally electing to spin in one direction. The dynamic impedance of the moving conductor then so changes the flux phase pattern that the conductor drifts from the centre and is finally ejected. This occurs in such a direction that the sphere appears to roll out of the field along an invisible horizontal plane.

Support of cylinders by elongated coil systems is also possible. Spheres and cylinders behave in an analogous manner to discs and rectangular plates as far as unresisted motion is concerned. A cylinder can be supported simply by a pair of

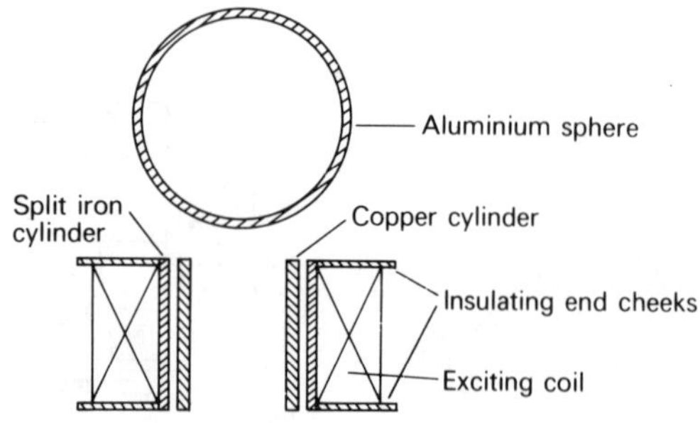

Fig. 4.10 Levitation of a sphere using shading ring stabilisers.

long conductors spaced horizontally and carrying high frequency currents in opposite directions. Again the single circular coil arrangement, used at high frequencies to support spheres, is analogous.

The techniques used in power frequency levitators are often useful in high frequency systems. The use of iron in the power frequency levitators is the same as in any conventional electric machine. Large numbers of conductors can be put in slots in the iron and yet they behave as if the current carried by the conductors in slots is concentrated at the top of the slots. Provided that the frequency is low enough to permit the use of iron, this results in considerable reduction of I^2R losses. It is now possible to extend the range of frequencies by the use of ferrites which have improved quite considerably.

The second way in which power frequency techniques are useful in the design of high frequency systems is the forms of construction. Highly rated coils or conductors such as water cooled tubes may be used in the same position as the conductors in the slots of Fig. 4.4, and this arrangement is an alternative to the conical coils of Fig. 4.13, which was adopted in the US. Finally the travelling field concept of power engineers is useful in deducing the best techniques for controlling the temperature of the suspended body. For a given configuration the pole pitch along one diameter, say, is fixed and increased frequencies will increase the velocity of the travelling field. This in turn will increase the power input to the supported member thus indicating that the temperature of the levitated member and the input required to lift a given mass may be controlled by varying the frequency. The power requirement per kg of lifted mass may be made as low as 130 W, although it must be borne in mind that the reactive power will still be quite substantial due to the inherently poor magnetic circuit for the primary.

A plot of lines of constant phase obtained experimentally[53] and as shown in Fig. 4.11 indicates that whilst a segment of the levitated sphere, between latitudes of 30° and 70° below the largest horizontal circle in this case, experiences an upward force the lower portions of the sphere experience a downward force. This downward force is experienced even by a conducting slab replacing the lower portion of the sphere. Thus the phenomenon of attraction of small non-ferrous conducting objects appears to be a feature of all levitation systems employing eddy current repulsion. Lovell[144] found that inside a cone such as that shown dotted in Fig. 4.12, small conducting objects experience a downward force, whilst outside the cone the force is upwards. It therefore appears that a conducting

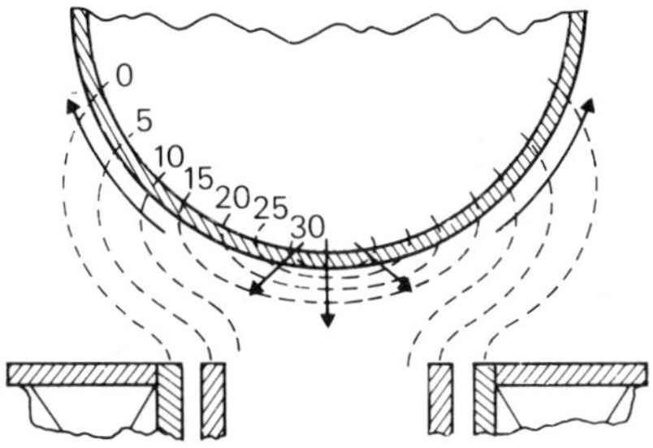

Fig. 4.11 Constant phase lines for a supported sphere (reproduced courtesy of the IEE).

sphere can only be levitated stably if it cuts this cone of attraction. Lovell proposed the use of this phenomenon for extracting from an eye small non-ferrous metal particles, employing the coil arrangement of Fig. 4.12. The attempt did not succeed because there appears to be a relationship between the force of attraction acting on such bodies, its change of temperature and its linear dimensions. A very small particle would in fact melt before it was accelerated due to the force of induction acting on it. In the case of proposed optical application the particle would cause a burn on the eye before the particle could be extracted.

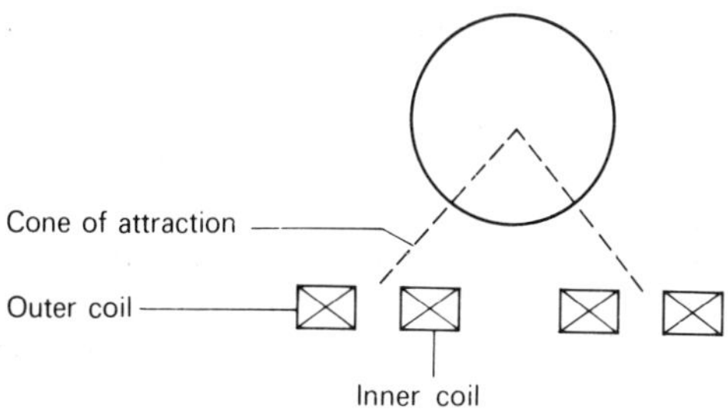

Fig. 4.12 Cone of attraction produced by a primary system (reproduced courtesy of the IEE).

The high frequency levitation work is mainly concerned with supporting molten metal, and laboratory equipment is available to achieve this. The equipment may be used for zone refining, and as this is achieved by a completely non-contacting method, i.e. without a crucible, it is claimed that the method leads to an extremely high degree of purity. The method generally appears[51,145–7] to be suitable for application to rather small masses of non-ferrous molten metal, although one experimenter claims to have suspended 8 kg of molten steel.[146] The coil arrangement of a typical piece of equipment is shown in Fig. 4.13. The conductors are copper tubes which are water cooled and thus highly rated, and the frequencies used are of the order of 10 kHz. The authors of reference 51 have attempted an analysis of the configuration of this figure by breaking down the system with single circular loops and calculating the axial force between each loop and a solid sphere placed on the axis of the loop. It is assumed that for a sphere placed in a uniform sinusoidally time varying field, not of a high frequency, the field due to the induced

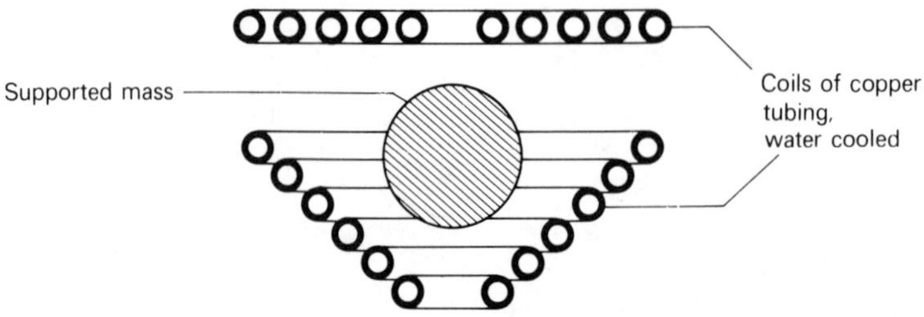

Fig. 4.13 High frequency system used by Orkress *et al.*

eddy currents is equivalent to that of a magnetic dipole alternating in time with the field of a phase and amplitude depending on the strength of the generating field, the radius of the sphere, its permeability and conductivity. The net force on the sphere is zero. If now the field is non-uniform in space the force on the sphere may be calculated by replacing it by a dipole whose moment is calculated as before in terms of a field whose properties are those of the field which exists at the centre of the sphere. The authors admit that this is only an approximation but the amount of detail in the resulting formulae is considerable.[53]

The coils shown in Fig. 4.13 consist of tubular conductors so that water may be circulated to assist in cooling them. With currents of the order of 800 A at 9.6 kHz, 250 g (0.5 lb) of bronze could be supported. It is estimated that about 50 kVA would be needed to support a few pounds of the metal.

The same system has been analysed more recently[148] for the examination of vertical stability. This is taken to mean the ability of the primary coils to produce resultant forces which oppose small axial displacement of the mass when supported at a given position. In spite of a detailed treatment of the topic, assumptions have to be made such as that the primary conductors are infinitely thin and the resulting formulae are even more complicated than those of reference 51.

4.3 Levitation of passenger carrying vehicles on the magnetic river

This concept is an extension only in some respects of the ideas of levitation of rectangular plates such as those described in Section 4.2 on p. 67. There are, however, fundamental differences when single sided motors and sheet secondaries with steel backing are used. The name magnetic river is intended to emphasise the behaviour of moving magnetic fields as analogous to viscous fluids. This is in experimental set-ups where rectangular plates suspended by one AC coil system are propelled by the same system. The moving magnetic field is regarded as a channel containing a flowing liquid into which objects can be dropped which thereafter are accelerated by the liquid and which if unrestrained soon attain the same speed as that of the liquid. If a piece of wood floats in the liquid stream of the 'river', and is caused to pull a mechanical load on wheels along the banks of the river as shown in Fig. 4.14, some water slips past the wooden block and the latter fails to reach the river speed V_s. Instead the load and the block will travel at some speed V and the entire force produced by the river on the piece of wood is transferred to the load on the banks via the connecting ropes.

There exists an analogy between the river and linear induction motors in which

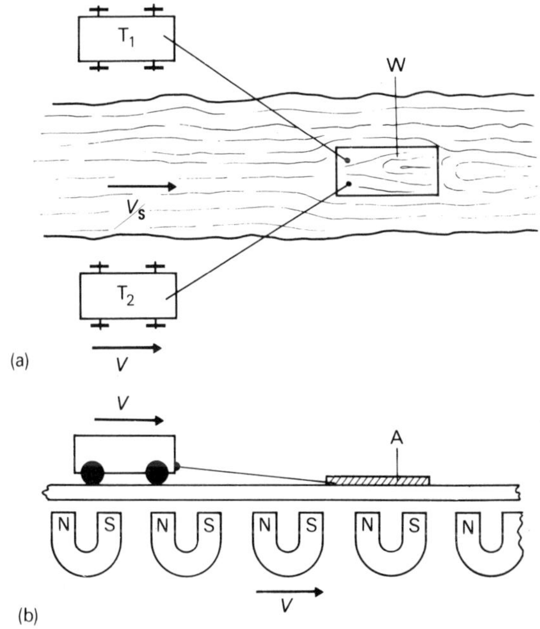

Fig. 4.14 Magnetic river analogy of linear induction motor: wooden block W in (a) is replaced by a piece of conducting material A in (b) (reproduced courtesy of the IEE).

the secondary member is repelled as well as propelled by the primary, and in the case of a machine in which the pole surface is horizontal causes the secondary to 'float' as does the wooden block in the river. It is possible to control the lateral, vertical, pitch and roll motions of the floating secondary sheet so that it is maintained in a horizontal plane within the confines of the electromagnetic river. The sheet must be made stable in yaw as well so that five degrees out of the six for a free body in space are under control. These are heave, roll, pitch, yaw and lateral displacement. The sixth degree of freedom is the linear motion along the stream or the direction of the travelling magnetic field.

4.3.1 Linear motors as propulsion and levitation devices

It is important to consider the limitations of linear motors which may be considered as the evolution of rotary induction motors cut along the axis and rolled out flat. Machines of this kind are called axial flux machines. The two essential elements of such machines, i.e. the magnetic circuit and the electric circuits, are shown in Fig. 4.15. When considered for application in high speed transportation,

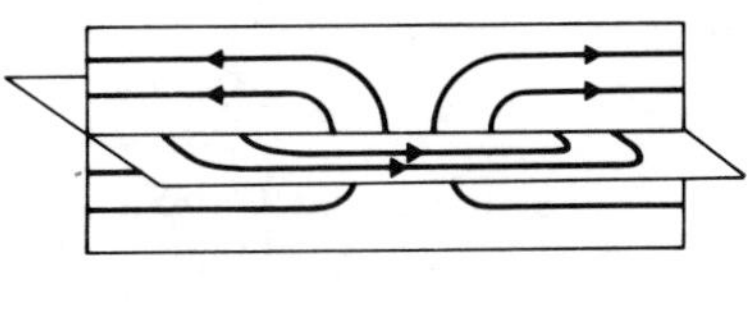

Fig. 4.15 Electric and magnetic planes related to the direction of motion for an axial flux machine.

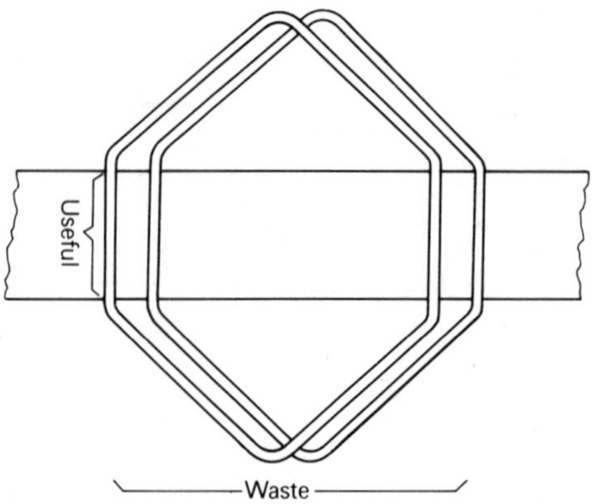

Fig. 4.16 Useful and wasted parts of the winding in a machine.

a long pole pitch is necessary, thereby stretching both the electric and magnetic circuit paths. This leads to large overhangs, i.e. wasted conductors on the windings (see Fig. 4.16), and an increase in the depth of the core required to carry the magnetic flux. This increases the weight of the machine, making it almost unacceptable for a transportation vehicle. Further, an axial flux motor cannot be used as a single sided machine acting against a reaction plate with backing steel (see Fig. 4.17(a)), without encountering considerable forces of attraction. These forces might almost double the weight of a levitated vehicle. Such forces of attraction are also present in conventional rotary machines, but if the airgap is constant around the periphery of the rotor these forces are self-cancelling. If not the result is unbalanced magnetic pull (UMP) which is taken up by the shaft bearings.

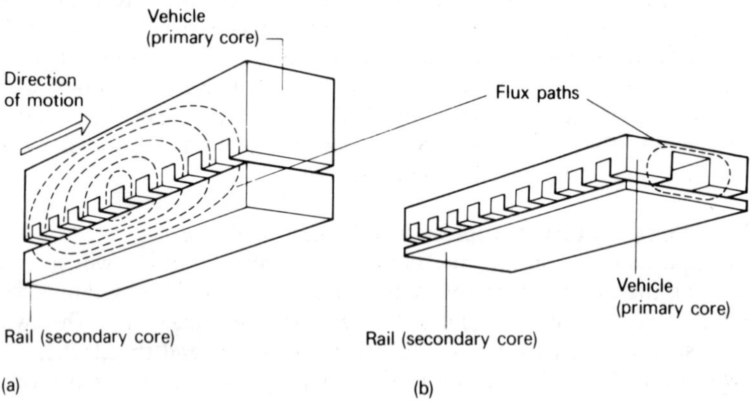

Fig. 4.17 Comparison of flux and relative thickness of magnetic circuits: (a) axial flux LIM; (b) transverse flux LIM.

These fundamental objections to axial flux motors in transport applications can be overcome by using a geometry called the transverse flux machine[54] (TFM); see Fig. 4.17(b). It can be seen from this figure that due to the pattern of flux in the core of the machine and the reaction rail which would normally be track mounted the thickness of the cores of both the primary and secondary would be dramatically reduced.

The principal advantage of the TFM becomes apparent when the equation for UMP is examined, particularly in relation to an electromagnetic arrangement where the goodness factor[127] G is high and where equal and opposite currents face each other across the airgap. The expression normally used for calculating the unbalanced magnetic pull is $B^2/2\mu_0$ which, although incomplete, is good enough in most cases of rotary machines because the force of repulsion between the opposite currents is less than 10% of the force of attraction between the magnetised surfaces. The full equation for the normal component unbalanced magnetic pull is[56]

$$F_n = \frac{B^2}{2\mu_0} - \frac{\mu_0 J^2}{2} \quad \text{per unit area} \tag{4.1}$$

In a typical linear motor design evolved straight from its rotary counterpart, the factor $B^2/2\mu_0$ will still dominate. However in a well-designed linear motor such as the one shown in Fig. 4.18, i.e. with wide slots and narrow teeth, the current density J may be increased by a factor of three and the flux density B reduced by a factor of 5, thereby making the second term much more dominant. The force is now not attractive but large and repulsive.

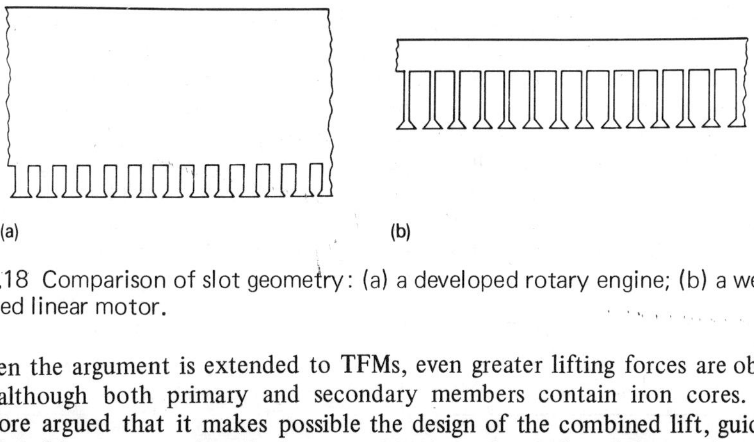

Fig. 4.18 Comparison of slot geometry: (a) a developed rotary engine; (b) a well designed linear motor.

When the argument is extended to TFMs, even greater lifting forces are obtainable, although both primary and secondary members contain iron cores. It is therefore argued that it makes possible the design of the combined lift, guidance and propulsion systems described as magnetic rivers. It was found during the course of development of TFMs which were primarily aimed at power factor improvement and track cost reduction that a primitive C-core TFM would levitate and stabilise any conducting sheet of sufficient thickness and width at any height within an expanding trough, as shown in Fig. 4.19. The power fed to the primary determines the height of levitation, and the stability at that height depends upon the levitated sheet being made to fit between the dotted lines shown in the figure.

By analogy with the unbalanced magnetic pull (UMP) of rotary machines the repulsion force obtainable from TFMs is called unbalanced magnetic push.[149] Further, the radial UMP is of the order of 20 times greater than the tangential or the useful force, and very little power, if any, is dissipated because there is no radial

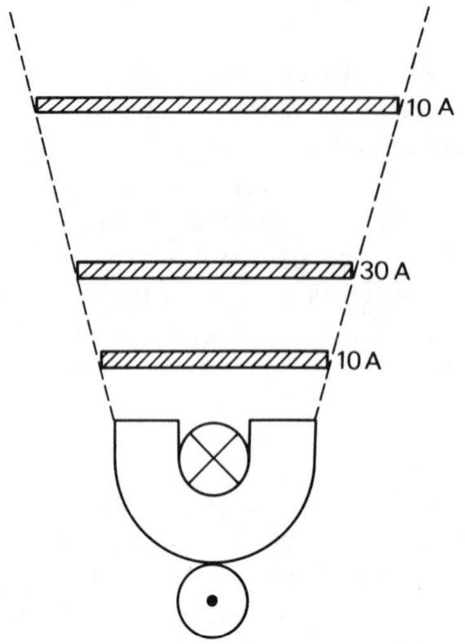

Fig. 4.19 Expanding geometry for stable levitation. Values of currents shown are those required in the primary coil to produce the same degree of stability at the heights shown (reproduced courtesy of the IEE).

motion and therefore no mechanical output produced. It is claimed that the levitation force is an unbalanced magnetic push exaggerated to its maximum by a single sided arrangement and is basically a force which does not require any power to levitate a mass. Whilst the above argument is true it does not take account of the fact that the higher the supported height, the lower the power factor due to secondary leakage reactance. Therefore the reactive voltampere requirements will be very substantial. The C-core TFM is stable and damped in heave, pitch, roll, yaw and lateral displacement when well designed[150] but has large primary leakage reactance due to windings along each side of the motor, while an E-core TFM is basically unstable in roll. The solution to this problem is believed to be a Xi-core[151] TFM (Fig. 4.20) where the leakage flux is contained by the side walls, without the loss of symmetry of the C-core.

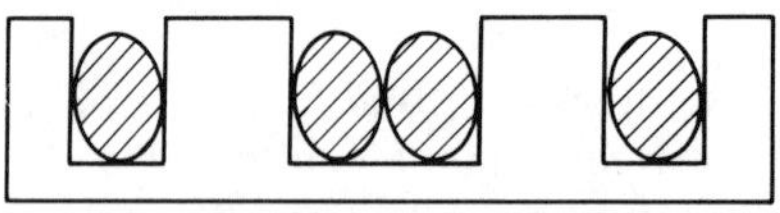

Fig. 4.20 Cross section through a Xi-core magnetic river.

The question of reactive power requirements of the magnetic river is examined in detail in Section 4.3.2 but it is claimed[152] that because of the implications of the ratio B_{gap}/J[143] and the fact that steel can be used to back up the aluminium track member, the bigger the machine the larger will be the induced current effects. These are compressive and stabilising whereas the magnetic pull effects which are tensile and destabilising will be reduced. This argument has often been used against the use of controlled DC electromagnets for levitation when talking about very large vehicles, such as those for carrying, say, 300 passengers and travelling at over 483 km h^{-1} (300 mph), perhaps weighing 100–150 tonne.

Early experiments showed that the normal force between the primary and secondary of a flat motor could be tensile or compressive at standstill, depending

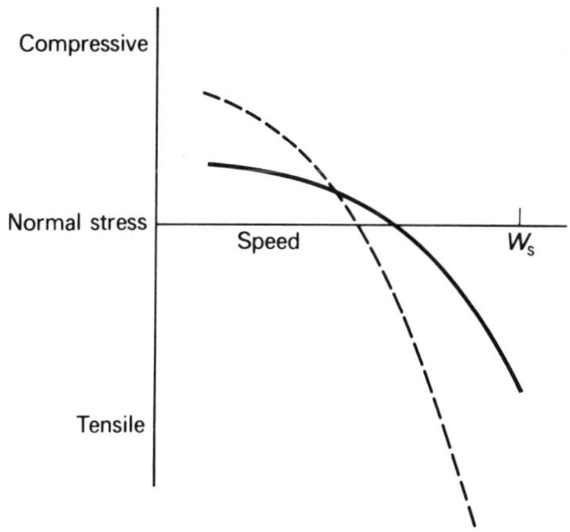

Fig. 4.21 Normal force variation with speed for a linear motor not designed with special regard for normal force.

upon the values of B_{gap} and J, but that as the motor gained speed, lift forces were reduced and tensile forces increased. A typical curve is shown in Fig. 4.21.

It was also observed that the front portion of a motor would exhibit a different normal force curve from that of the rear portion shown dotted in Fig. 4.21.[153] This could give rise to a dolphin effect or pitch instability and it was found that increased goodness factor G would aggravate such instability. The normal force–speed curves given in reference 153 show that the larger the airgap the flatter the normal force–speed curve (see Fig. 4.22). Use of large airgap g or large pole pitch p, in order to maintain a high value of the goodness factor $G = p^2\mu_0\omega/\pi^2\rho_r g$, both work in favour of the concept of magnetic river, using TFMs exploiting their short magnetic

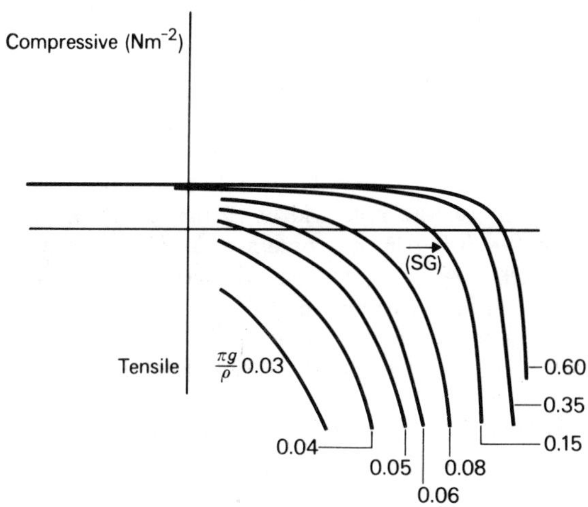

Fig. 4.22 Variation of normal force–speed characteristics with airgap length (reproduced courtesy of the IEE).

circuits (see equation (4.3)). This means that magnetic river is an all or nothing exercise.[136]

4.3.2 Magnetic river as a vehicle system

Some theoretical boundaries of this system have been calculated by Eastham.[58] The reactive input to the machine, considering only the stored energy in the tangential support field, is given by

$$\text{VAR} = 2F_n g\omega \tag{4.2}$$

Thus for a 50 tonne vehicle energised from a 50 Hz supply at 0.1 m gap the minimum requirement is 30.8 MVAR. The suspension is soft and 0.1 m is thought to be the minimum possible without some form of feedback height control. The MVAR boundary is absolute and applicable to any form of suspension which uses an AC field. The stator leakage will increase this minimum MVAR requirement even further. A vehicle of this size will require 3.25 MW for it to be driven at $120\,\text{m s}^{-1}$.

Another limiting factor in design can be discovered by considering that a machine of this type will have a high rotor-coupled time constant ($\tau = p^2\mu_0/\pi^2\rho_r g$) due to the large pole pitch p. Peak force for such machines is given by [154]

$$F_T = \frac{n'\rho_r J_s^2}{v_s} \times \text{area} \tag{4.3}$$

where n' = number of poles n plus 1; ρ_r = rotor surface resistivity; J_s = current loading; and v_s = synchronous speed.

At constant J_s this peak force will occur at a slip of $1/(n+1)$ and the peak efficiency is $n/(n+1)$ excluding the stator loss. It is assumed that the normal lifting force for the high rotor time constant cases is given by

$$F_n = \frac{\mu_0 J_s^2}{2} \times \text{area} \tag{4.4}$$

A machine with four poles (for compactness) and a pole pitch of 1.5 m will give a synchronous speed of $150\,\text{m s}^{-1}$ and the peak force and efficiency will occur at $120\,\text{m s}^{-1}$. The machine will be 6 cm long. Two such machines, one at each end, will be employed because the stators have pitching effects as well as normal forces at low slips. The limit on the pole widths is then about 0.5 m, yielding a U-core machine about 2 m wide. Putting the pole area at $12\,\text{m}^2$, in order to obtain $F_n = 50 \times 10^3 \times 9.81$ N, the required current density from equation (4.4) is $J = 2.55 \times 10^5\,\text{A m}^{-1}$. Such high current can only be achieved by direct water cooling. This therefore sets a fairly rigorous limitation to the design unless more area is found but the machines will occupy a considerable amount of the floor area of the vehicle as it is.

Dividing equation (4.3) by equation (4.4) we obtain

$$\frac{F_n}{F_T} = \frac{v_s\mu_0}{2n'\rho_r} \tag{4.5}$$

One and only one value of ρ_r will set the ratio of F_n to F_T equal to the weight-to-drag ratio of the vehicle. Inserting values of $F_n = 490$ kN and $F_T = 27.1$ kN, an aluminium plate of approximately 30 mm thickness is required. Since this is far greater than the skin depth at 50 Hz the effective resistivity will be high at starting conditions but will be reduced to the correct value as the speed increases. The machines which emerge from this simple analysis have a large total area but still a high current loading, and Eastham estimated that maximum possible efficiency will be 20% and the best possible full load airgap power factor will be 0.138.

4.4 Summary

Eastham has presented the two designs outlined in Table 4.1, one for 20 mm gap and one for the 100 mm gap. These designs are based on programmes developed from those used for previous work[153,155] and are founded on the layer theory.

Table 4.1

Vehicle weight 50 tonne	20 mm	100 mm
Motor weight	11.2 tonne	11.2 tonne
Sync. capacitor weight	19 tonne	29.5 tonne
Corrected power factor	1.0	1.0
Peak efficiency	0.38	0.38
Weight remaining for payload and vehicle body	19.8 tonne	9.3 tonne

The vehicle will not operate without feedback control at a gap of 20 mm and therefore these predictions are disappointing as well as counter to the arguments used against the controlled DC electromagnet systems. The figures given above may be slightly pessimistic but some of these have been verified and there is little reason to believe that they are likely to be widely out. Clearly any proposals for the use of magnetic river need to be preceded by a feasibility study using the best available techniques to optimise the design. The use of magnetic river design for high lateral to normal force ratios, however, puts them high in the list of normal transverse flux machines with the possibility of use as propulsion units for high speed levitated vehicles.

5

Suspension using controlled DC electromagnets

5.1 Introduction

The most advanced technology of suspension at the time of writing this book is that using controlled DC electromagnets. This technology was developed during the 1970s in Germany, Japan, the US and the UK, and vehicles of various sizes, weighing between 1 tonne and 35 tonne, have been operated successfully in these countries, at speeds ranging from a few kilometres per hour to over 400 km h^{-1}. There are also ambitious plans for the next few years, for example in Japan in 1978, a 2.3 tonne, 9 passenger, 7 m long vehicle was built to operate at a speed of 100 km h^{-1} (see Fig. 5.1). The predecessor of this vehicle, a 1 tonne vehicle, was tested in February 1978 at a speed of 307.8 km h^{-1}. There are currently designs underway of an 80 seat pre-operational test vehicle which is expected to start undergoing tests in the early 1980s. The speed of this vehicle will be 300 km h^{-1}.

Fig. 5.1 Japan Air Lines 8 passenger vehicle (HSST-02) using controlled DC electromagnets for suspension and linear motor for propulsion (photograph courtesy of Japan Air Lines).

Fig. 5.2 MBB 36 tonne magnetically suspended vehicle using linear synchronous motors for propulsion. The stator windings are on the guideway and the lift magnets also act as the excitation magnets for the motor. The vehicle was demonstrated in the Hamburg exhibition in the summer of 1979 (photograph courtesy of MBB).

In Germany, since 1969, six vehicles have been built and tested. In June 1979 a 68 passenger vehicle weighing 36 tonne (see Fig. 5.2) was demonstrated at the Transport Exhibition in Hamburg. This is a precursor of a 100 passenger, 121 tonne vehicle which is under construction and which is to be operated on a 31.5 km track in Emsland between Meppen and Papenburg (see Fig. 5.3).

The Japanese vehicle HSST-02 is propelled by a linear induction motor mounted on the vehicle but the German vehicle Transrapid 05, which was demonstrated on a 900 m track, is a long stator machine, i.e. the stator winding is on the track. Further, the machine is a synchronous linear motor which uses the lift magnets as excitation magnets (see Fig. 5.4). It is proposed that the 121 tonne vehicle will have an aircored linear synchronous motor with windings on the track and perhaps superconducting excitation magnets. This scheme therefore eliminates the need to collect power through sliding brush contacts at high speeds. A novel feature of the German vehicles is that the magnets are mounted on individually sprung members, and this concept has been labelled the 'magnet wheel'. A variant of this, a sprung chassis similar to that in motor cars, has been used by Japan Air Lines for their HSST-02 vehicle, by General Motors and by the author at the University of Sussex.

Apart from the achievements described above in the vehicle area, considerable work is in progress in applying controlled DC electromagnets to contactless or frictionless supports in place of conventional bearings. Some of these are described in Section 5.6.

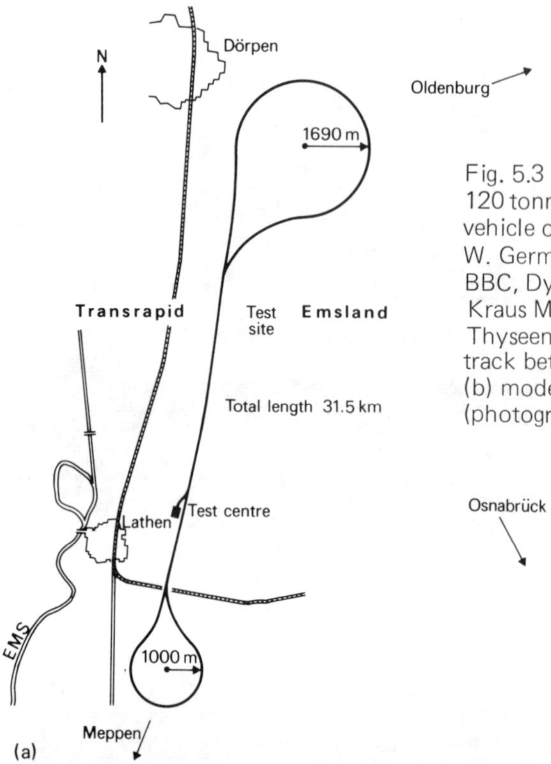

(a)

Fig. 5.3 Proposed scheme for a 120 tonne magnetically suspended vehicle operating in Emsland, W. Germany, to be fabricated by AEG, BBC, Dyckerhoff and Widmann, Kraus Maffei, MBB, Siemens, and Thyseen Henschel by 1982. (a) 31.5 km track between Meppen and Dörpen; (b) model of the proposed vehicle (photograph courtesy of MBB).

(b)

Fig. 5.4 Detail of the Hamburg exhibition vehicle showing the suspension and guidance magnets as well as the track windings of the LSM (photograph courtesy of MBB).

5.2 Principle of suspension using controlled DC electromagnets

A corollary of Earnshaw's theorem[2] and Braunbeck's subsequent work[3] is that systems using permanent magnets or electromagnets (AC or DC) without control of current are inherently unstable. In order to achieve stable suspension it is necessary to devise a means of regulating the current in an electromagnet using position feedback of the object to be suspended. The effect of this is to modify the force–distance characteristics such that the current and thus the force of attraction decrease as the gap decreases and vice versa. A simple method of detecting changes in gap is the photo-optical sensing method which is illustrated in Fig. 5.5(a). If the steel ball to be suspended in this case is attracted towards the magnet, the amount of light falling on the photocell (phototransistor) diminishes, and this in turn decreases the coil current and the force acting against gravity. The force–distance characteristics are therefore modified as shown in Fig. 5.5(c). In order to use this

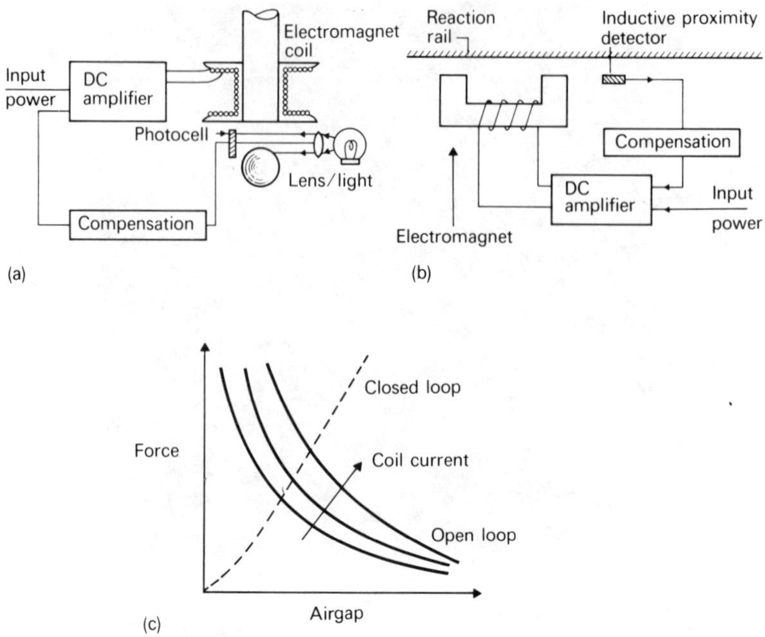

Fig. 5.5 Principles of magnetic suspension using controlled DC electromagnets: (a) scheme of suspension using an optical transducer; (b) suspension of a controlled electromagnet under a steel rail; (c) open and closed loop force–distance characteristics.

method of suspension for bodies such as vehicles, the magnets and the amplifiers must be mounted on the moving member (the vehicle) and the optical transducer is replaced by something more appropriate to the application. The modifications required are shown in Fig. 5.5(b), where the magnet is now suspended below an inert rail and the transducer is labelled 'inductive proximity transducer'. This is only one of several methods which may be used in this application. The system described here is a closed loop position control system, and hence at least some form of anticipation of position change is required in the feedback path. This is shown in the figure as compensation. The simplest form of compensation is derivative of the position signal, but with more complex and multimagnet systems, other forms of compensation as well as transducers become necessary if accurate stability is to be achieved. Fig. 5.6(a) shows the suspension of a 60 g steel ball, using the optical transducer,[156] and the circuit diagram of the suspension system is given in Fig. 5.6(b). The circuit shows the lead compensation to the signal from the phototransistor.

5.2.1 Nature of the control problem in a single magnet system

The force of attraction between magnetised bodies is given by

$$F_m = \frac{B^2}{2\mu_0} \times \text{area} \tag{5.1}$$

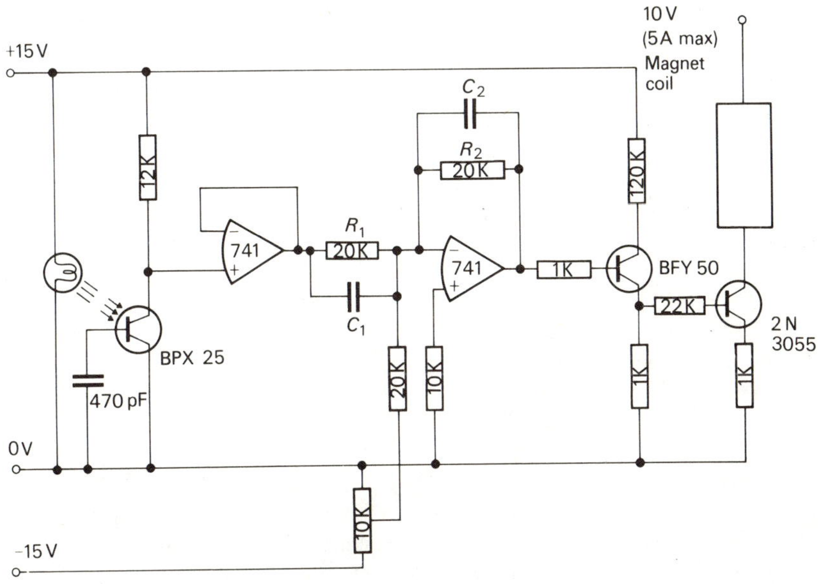

(b)

Fig. 5.6 Experimental suspension models: (a) suspension of a steel ball using optical transducer; (b) circuit for the steel ball suspension, i.e. single magnet suspension.

In the case of electromagnets such as those used in suspension systems operating with an airgap, the gap flux density is directly proportional to the ampere turns NI and inversely proportional to the gap length g, i.e.

$$F_{\text{m}} = \frac{1}{2\mu_0}\left(\mu_0 \frac{NI}{z}\right)^2 \times \text{area} \tag{5.2}$$

It is obvious from this equation as well as the force–distance characteristics of Fig. 5.5(c) that the system is a highly nonlinear position regulator. However, an adequate insight into the nature of the control problem can be gained by looking at a linearised model. At the nominal required gap in Fig. 5.5(b) the magnet current i_0 generates a force equal to the weight to be suspended. Any displacement from this position results in a change i in the current. It is assumed in the first instance that the change in the force of attraction f is given by a linear function of the gap z and changes of current i. The linearised equation is therefore

$$f = -k_1 z + k_2 i \tag{5.3}$$

for small changes about the equilibrium position, where k_1 is the force per metre at constant current and k_2 is the force per ampere at constant distance. The constants k_1 and k_2 can be determined experimentally for a given magnet. Thus

$$m\ddot{z} = -f = -(-k_1 z + k_2 i) \tag{5.4}$$

where m is the mass of the suspended body.

The relationship between the magnet current and the voltage is

$$Ri + L\frac{\text{d}i}{\text{d}t} = V \tag{5.5}$$

where R is the resistance and L the inductance of the magnet coil.

In Laplace transform notation equations (5.4) and (5.5) become

$$\left.\begin{aligned} \left(s^2 - \frac{k_1}{m}\right)Z(s) &= \frac{k_2}{m}I(s) \\ \text{and} \qquad I(s) &= \frac{k}{(1 + sT_{\text{m}})}V(s) \end{aligned}\right\} \tag{5.6}$$

Combining the two equations (5.6) gives

$$\frac{Z(s)}{V(s)} = \frac{kk_2}{m(1 + sT_{\text{m}})(s^2 - k_1/m)} \tag{5.7}$$

The quantity that can be directly controlled through a DC amplifier driving the magnet is its voltage output, and T_{m} represents the lag in the resulting current due largely to the magnet inductance. However, the second term $(s^2 - k_1/m)$ arises from the force–distance characteristic, which may be considered analogous to a spring stiffness constant, but in this case negative. The consequence of the negative spring stiffness, characterised by k_1, causes these systems to be basically unstable. The explanation in control terminology is that the term $(s^2 - k_1/m)$ represents poles at $\pm\sqrt{k_1/m}$ on the s-plane and the pole at $+\sqrt{k_1/m}$ contributes to the instability. In the closed loop or feedback system shown in Fig. 5.5, the transducer will have a very small time constant associated with it but it is of very little practical significance here in comparison with the main time constant. Because the

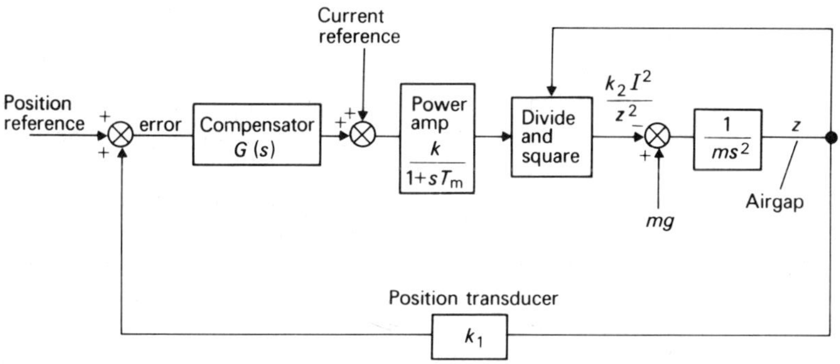

(a)

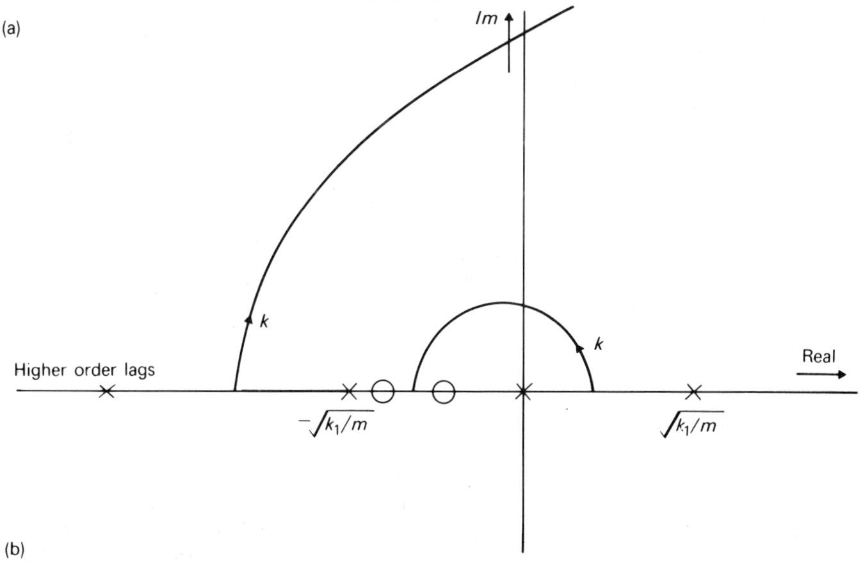

(b)

Fig. 5.7 Magnetic suspension with cascade compensation: (a) block schematic of the controller; (b) root locus showing the principal poles and zeros.

compensation is in series with the transducer or the actuating signal and the amplifier driving the magnet, this scheme is known as series compensation or cascade compensation. A block schematic of the system and the corresponding root locus diagram are shown in Fig. 5.7.

5.2.2 Multimagnet systems

In multimagnet systems and particularly in transport applications there are some departures from the single magnet systems. Generally, when providing maximum lift the magnet is operating in a highly nonlinear part of its force–distance characteristics (see Fig. 5.5(c)). The control system gain is directly related to the slope of the characteristics at the operating point. Therefore changes in the operating conditions, such as load variations, will severely degrade the transient response or even cause instability. Another important departure of multimagnet systems arises from the mechanical coupling of magnets mounted on a rigid chassis and that of the individual control systems of the magnets. This demands an even higher degree of stability than that obtainable from the cascade compensation scheme using position transducers alone.

There are several ways of overcoming, or at least reducing, the severity of the problem of mechanical interaction. The flexible chassis or magnetic wheel has already been mentioned in Section 5.1, and is described in greater detail in Section 5.3.2. Although this results in enhanced stability, much improvement is also needed in the design of the basic control systems, to increase the stability further. If a transducer is introduced on the face of the lift magnets to measure the gap flux density, and used in conjunction with a feedback loop to control the magnet current in such a manner that the gap flux density remains constant over the operating range of airgaps, the nonlinear force–distance characteristics will be very nearly linear. A primary cause of instability will therefore be eliminated, at least in the operating range, by the use of a separate flux control loop. Although other transducers have variously been used, the author has used a Hall effect device mounted on the face of the magnets for the simple reason that the output from a Hall plate is DC.

The use of a flux control loop means that the force is now independent of the gap length or in equation (5.3) the constant k_1 is zero. Alternatively, this also means that in the s-plane the pole in the right hand half at $+\sqrt{k_1/m}$ will have effectively moved to the origin. This partial linearisation, leaving the force variations proportional only to i^2, makes a dramatic improvement to the stability margin of multimagnet systems. Introduction of this loop to keep airgap flux density constant changes the open loop system equation (5.7) to

$$\frac{Z(s)}{V(s)} = \frac{kk_2}{m(1 + sT_{\mathrm{m}})s^2} \tag{5.8}$$

The concept of using an independent feedback loop for one parameter, gap flux density in this case, can be extended to the independent measurement of position and velocity and to using separate loops for each. This scheme of partial state feedback gives control over gains of the individual loops and enables optimum setting for each. The transducers employed for measurement of position and velocity are described later but measurement of acceleration and integrating the accelerometer signal to give velocity has the advantage that noise is attenuated in the process of integration. In obtaining velocity from the position signal by taking its derivative or by differentiation introduces noise. If the position signal is obtained by a process of rectification of an AC voltage, the problem can then be acute. A block schematic of a single magnet system using separate position, velocity and flux transducers is shown in Fig. 5.8.

An efficient suspension system employing DC electromagnets must necessarily operate at relatively small airgaps compared to, say, the superconducting magnet levitation system. Control of the airgap must therefore be very tight, and for passenger carrying vehicles this introduces conflicting requirements. Tight control of airgap or a stiff suspension will produce a hard and uncomfortable ride. It is thus commonly accepted that acceleration feedback is also necessary to achieve ride control.

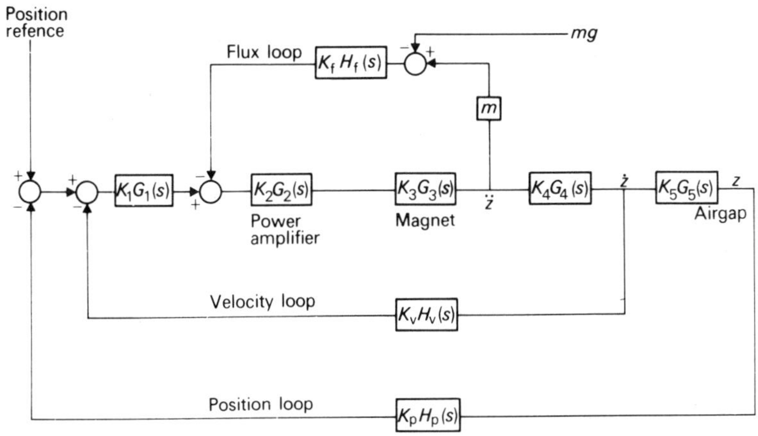

Fig. 5.8 Schematic of controller with partial state feedback.

5.3 Analytical aspects of multimagnet systems

The difficulties of stabilising and extending the analysis of single magnet systems to multimagnet systems are many and formidable. As pointed out in Section 5.2 they arise, in the main, through the mechanical cross coupling between the magnets and their control systems on a rigid or flexible chassis. For a four-magnet chassis an analogy may be drawn with a four-legged stool standing on an uneven floor. The attitude of the chassis is determined by heave, roll and pitch, i.e. three parameters. Specifying and attempting to control four gaps such that they are the same therefore introduces a conflict between the control loops for each corner. The chassis is seldom absolutely rigid and hence there will be substantial torsional deformation and stresses set up within the chassis. This method of control of individual gaps is equivalent to local control. Alternatively, the measurement of four gaps can be converted to give the attitude of the vehicle or chassis in terms of heave, roll and pitch; these may be controlled instead, by associating a reference as well as compensation with each of them. This may be termed integrated control. The concepts of local and integrated control are illustrated in Fig. 5.9. This figure illustrates the problem which is that multimagnet vehicle frames are in essence nonlinear multivariable control systems and that the object is to achieve decoupling by whatever means are practicable. These and various other aspects of control of magnetically suspended vehicles, such as operation of flexible guideways, received a great deal of attention during the 1970s.[92–5,157–67] It should be noted that the analysis given here in outline is basically applicable to the dynamics of vehicles using other methods of levitation, say superconducting levitation, by varying principal suspension parameters such as stiffness, damping, etc.

5.3.1 Equations of motion and geometric transformations[168]

A free body in space has six degrees of freedom: three associated with translational motion, the other three with rotation. Given information about all external forces acting on the body, the principles of linear and angular momentum can be used to derive equations of motion. Generally six second-order nonlinear differential

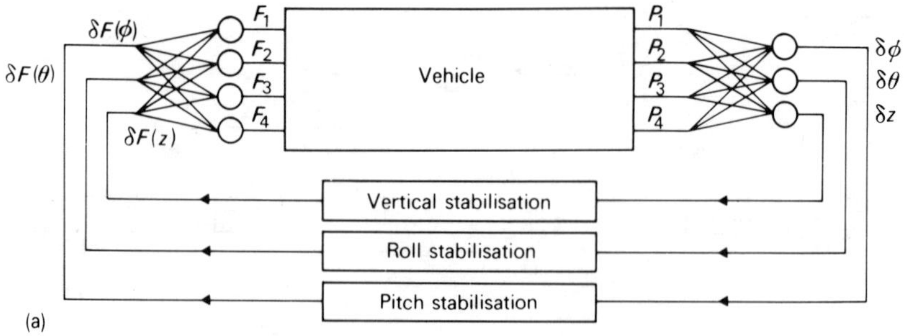

(a)

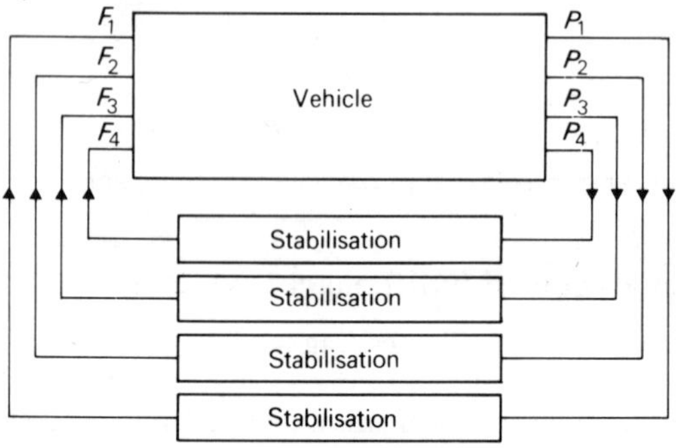

(b)

Fig. 5.9 Multiple magnet suspension schemes: (a) control of each degree of freedom; (b) control of each magnet—transducer pair.

equations are obtained which contain common factors giving rise to interaction between the variables.

Consider the vehicle to be represented by a uniform box shaped object with the centre of mass coincident with the centre of geometry (see Fig. 5.10). The frame of reference and the principal axes are as shown in the figure. The principle of linear momentum leads to the following equations:

$$m\ddot{x} = F_x \qquad m\ddot{y} = F_y \qquad m\ddot{z} = F_z \tag{5.9}$$

where F_x, F_y and F_z are the resultant forces acting along the inertial axes $0x$, $0y$ and $0z$.

The principle of angular momentum leads to three more torque equations for the rotational coordinates. From Euler's equations, after substitution of appropriate Euler's angles, we get

$$\begin{aligned} T_x &= I_{xx}\dot{\omega}_x + (I_{zz} - I_{yy})\omega_y\omega_z \\ T_y &= I_{yy}\dot{\omega}_y + (I_{xx} - I_{zz})\omega_z\omega_x \\ T_z &= I_{zz}\dot{\omega}_z + (I_{yy} - I_{xx})\omega_x\omega_y \end{aligned} \tag{5.10}$$

A suitable set of Euler's angles is shown in Fig. 5.10, which corresponds to the notion of rolling, pitching and yawing of the vehicle frame. In terms of these angles, the angular velocities are

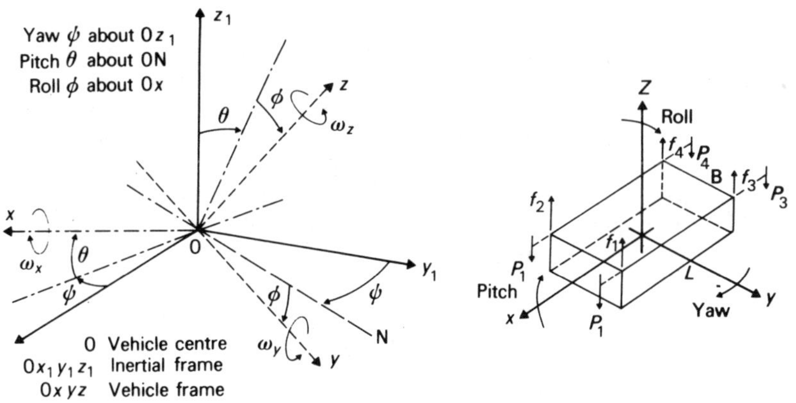

Fig. 5.10 Frame of reference and dimensions used in equations of motion. 0x, propulsion; 0y, lateral axis; 0z, heave axis. Quantities L, B: distances between centres of lift along x- and y-axes respectively. Quantities 2l and 2b: distances between position transducers along x- and y-axes respectively. Quantity 2d: diagonal distance between position transducers. Quantity D: diagonal distance between centres of lift magnets.

$$\begin{aligned}\omega_x &= \dot{\phi} + \dot{\psi}\cos\theta \\ \omega_y &= \dot{\theta}\sin\phi - \dot{\psi}\sin\theta\cos\phi \\ \omega_z &= \dot{\theta}\cos\phi + \dot{\psi}\sin\theta\sin\phi\end{aligned} \tag{5.11a}$$

and

$$\begin{aligned}\dot{\omega}_x &= \ddot{\phi} + (\ddot{\psi}\theta + \dot{\psi}\dot{\theta}) \\ \dot{\omega}_y &= \ddot{\theta} + (\dot{\psi}\dot{\theta}\phi\theta - \ddot{\psi}\phi - \dot{\psi}\phi - \dot{\theta}\dot{\phi}\phi) \\ \dot{\omega}_z &= \ddot{\psi} + (\ddot{\theta}\phi + \dot{\theta}\dot{\phi} - \dot{\psi}\dot{\phi}\phi - \dot{\psi}\dot{\theta}\theta)\end{aligned} \tag{5.11b}$$

For small angular motions the terms in brackets are negligible compared to the principal terms $\ddot{\phi}$, $\ddot{\theta}$ and $\ddot{\psi}$, i.e. in comparison to roll, pitch and yaw accelerations. For example, a vehicle with magnets 1 m apart working at nominal gaps of 10 mm ± 2 mm would have corresponding angular displacements of the order of 5 milliradians.[168] The resulting equations for angular motions and torques therefore become

$$T_x = I_{xx}\ddot{\phi} \text{ (roll)} \qquad T_y = I_{yy}\ddot{\theta} \text{ (pitch) and } T_z = I_{zz}\ddot{\psi} \tag{5.12a}$$

where T_x, T_y and T_z are the applied moments of the external magnet forces. In practice, the platform may not be absolutely rigid and hence one more equation describing the torsional motion of the platform is necessary. Again, a linearised expression of this motion is

$$T_t = I_t\ddot{\xi} \tag{5.12b}$$

Equations (5.9) and (5.12) describe the vehicle dynamics for the case assumed. Modifications are necessary to account for variations such as the centre of mass, the origin of coordinates and the centre of geometry not being coincident.[166]

The equations can still be linearised but are more complex. In practice the forces and torques in equations (5.9) are derived from a system of electromagnets, and the displacements and rotations are measured by a system of transducers. Since the force, torques, displacements and rotations are not measured directly in the form in which they occur in equations (5.9) and (5.12), the magnets and transducer systems effectively introduce cross coupling. The four lift forces generated by the corner magnets, f_1, f_2, f_3 and f_4, are controlled by the outputs p_1, p_2, p_3, p_4 from a set of four transducers so as to maintain the roll and pitch angles at zero and the height of the vehicle constant. Then in Fig. 5.11,

$$\begin{aligned} F_z &= \tfrac{1}{4}(f_1 + f_2 + f_3 + f_4) \\ T_x &= (-f_1 + f_2 - f_3 + f_4)B \\ T_y &= (f_1 + f_2 - f_3 - f_4)L \end{aligned} \tag{5.13a}$$

and

$$\begin{aligned} z &= \tfrac{1}{4}(z_1 + z_3 + z_4) \qquad \phi = \frac{1}{2b}\sum_{i=1}^{4} \pm z_i \\ \theta &= \frac{1}{2}\sum_{i=1}^{4} \pm z_i \qquad \psi = \frac{1}{2d}\sum_{i=1}^{4} \pm z_i \end{aligned} \tag{5.13b}$$

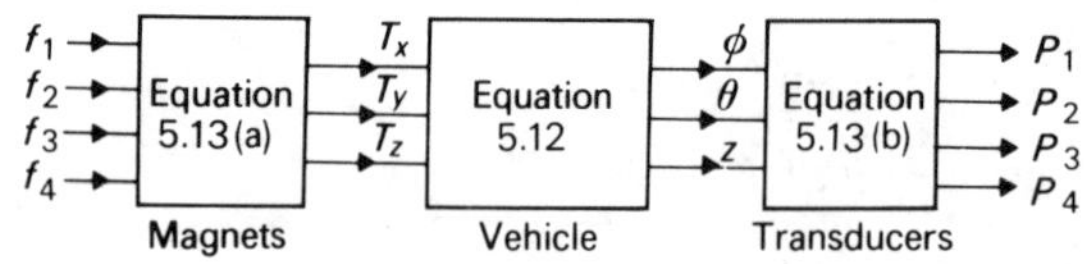

Fig. 5.11 Block schematic of the dynamics of a four-magnet vehicle.

Thus the vehicle can be regarded as a transformation between the magnet forces and the transducer outputs and is in effect a four input, four output system. Change in any one of the forces will cause a response in all the output variables in view of the cross coupling implied by equations (5.13). However, the essentially noninteracting dynamics represented by equation (5.12) enables in principle each degree of freedom ϕ, θ and z to be controlled independently by working with the appropriate linear combination of the f_i and p_i. For example, to control the vertical coordinate z an equal component in f_i is set aside for this purpose and is made to depend on the mean value of p_i. A typical control loop would then appear as in Fig. 5.9(a).

An alternative design procedure is to admit the cross coupling implied in equations (5.13) and to attempt control based on a tight association between a magnet and its nearest transducer, i.e. the force f_i is controlled by the transducer p_i. This leads to the control loop of the form shown in Fig. 5.9(b). Clearly in this case disturbances entering any loop or applied to the mechanics in the form of external forces will cause changes to all variables and will propagate throughout the system. This can be compared, for example, with the effect of a vertical force applied at the centre of mass which will cause a change in z only for the former type of controller. Responses for step change in airgap reference for the two modes of control are shown in Fig. 5.12. The response for torsion in the chassis

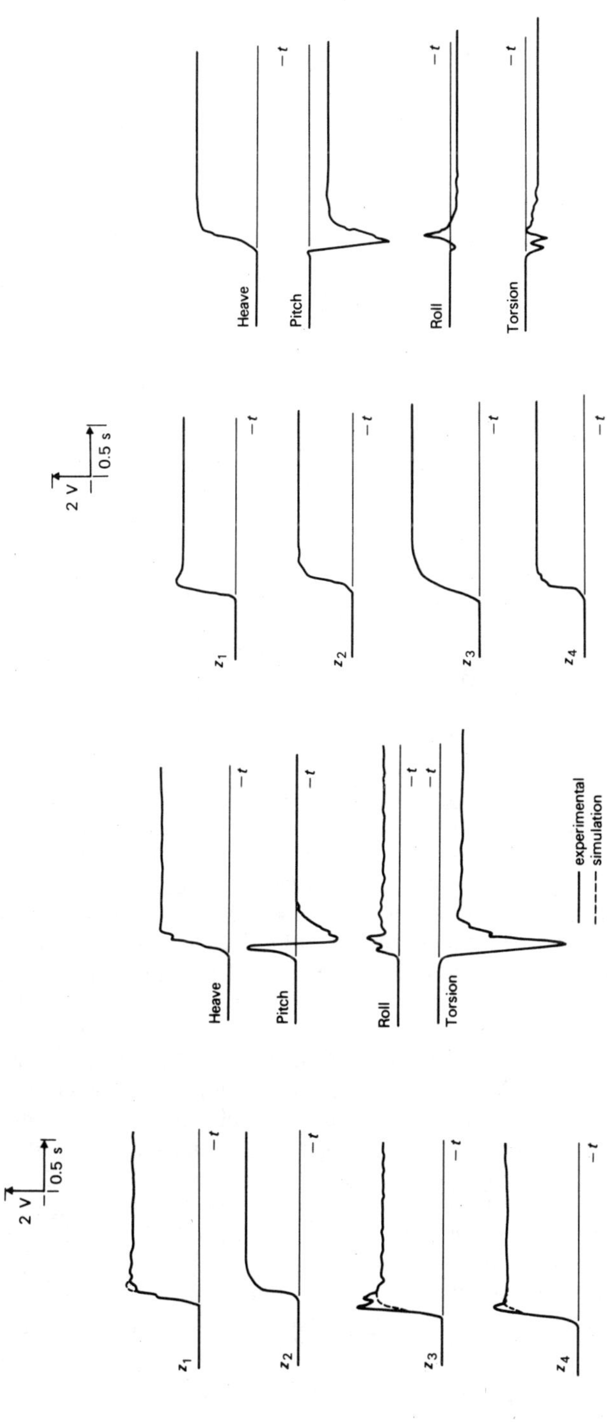

Fig. 5.12 Responses for step changes in airgap reference input of each corner: (a) with local control; (b) with integrated control.

is particularly worth noticing. Whilst the arrangement which is interacting presents problems in analysis, it is physically simpler than the one which is arranged such that each degree of freedom is controlled independently. It must be borne in mind that all magnets and transducer pairs are assumed in the preceding discussion to be identical, a state of affairs seldom likely to be achieved in practice. There will therefore always be interaction, both static and dynamic, in multimagnet vehicles. Further, although static noninteraction may be very nearly achieved, complete dynamic noninteraction under all operating conditions is a virtual impossibility, partly because of the higher order effects indicated but also because the centre of mass is different from the centre of geometry, and this introduces coupling between various modes. The problem of control of multimagnet vehicles gives a very good insight into the control of multivariable nonlinear systems.

5.3.2 Decoupling through flexible chassis

A somewhat simpler alternative to dynamic decoupling by the use of state feedback is to introduce flexibility in the vehicle chassis. The advantage of this method is that it does not require any electronic circuitry other than that for individual magnet stabilisation. This scheme was implemented by the author in one of the vehicles (see Fig. 5.13). In this vehicle platform, the magnets were mounted as a pair on beams which are pivoted about their centres and coupled to the vehicle platform by means of springs of appropriate stiffness. The nearest equivalent to this arrangement is the independent suspension on a motor car. The responses of such an arrangement to sinusoidal disturbances are shown in Fig. 5.14 and this illustrates how much has been achieved. The same principle of a sprung chassis has been utilised in the Japanese vehicle HSST-02 and the General Motors vehicle.

The concept of independent control of magnets and sprung chassis has been developed even further on the German vehicles, Transrapid 04 and Transrapid 05. In these vehicles the magnets are attached via a primary suspension to magnet frames (see Fig. 5.15(a)) which are connected to the cabin by a secondary suspension. This structure adapts to guideway irregularities even at high speeds and therefore a smaller nominal airgap may be realised, leading to a considerable reduction in magnet weight and power required for suspension. The mechanical details of the mounting of magnets are shown in Fig. 5.16. The frequency responses and results of other tests indicate that the modular structure, as shown in Fig. 5.15, has profound influence[93,162] on the ability of the vehicle to cope with guideway irregularities at high speeds with reduced gaps and also on other factors such as the induced eddy current losses and thus the overall suspension power, ride (i.e. cabin acceleration), etc. Magnets are supplied by their own magnet drivers. When these current drivers are supplemented by control networks, individual sensors and individual power supplies, this forms the basic subsystem of the decentralised control system. This is called the single magnet controller. The configuration may be considered for on-board computer purposes to be a decentralised hierarchical system of two levels. The controlled magnets form level 1 and this level has to provide system stability under all operational conditions even in the case of total failure of level 2 or partial failure of level 1. Functions of level 2 are bias adjustments, control of take off and landing, system self check and on-line monitoring. When decentralised hierarchical control is applied to suspend and guide a modular structure vehicle, the unit consisting of a single magnet controller and single magnet

(a)

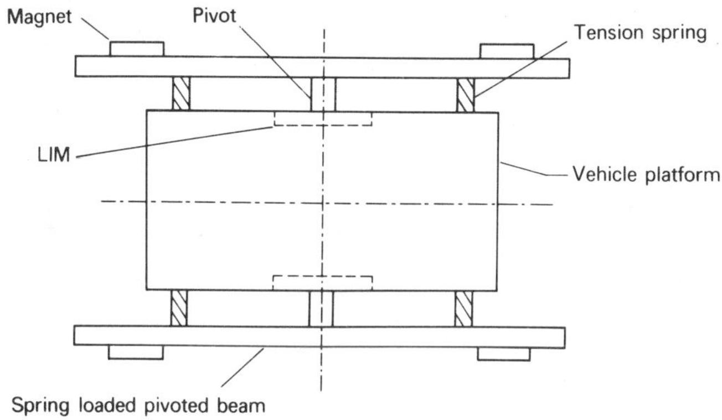

(b)

Fig. 5.13 Flexible chassis vehicle: (a) an experimental 500 kg vehicle; (b) schematic of the vehicle.

with its mechanical suspension elements is sometimes called a 'magnetic wheel'. To design the single magnet wheel controller only the magnet wheel is considered under the simplifying assumption of being attached to a frame of much bigger mass.

5.4 Transducers, magnets and power amplifiers

The suspension system using controlled DC electromagnets is in essence a position regulator or a control system and therefore the principal elements of it, i.e. the

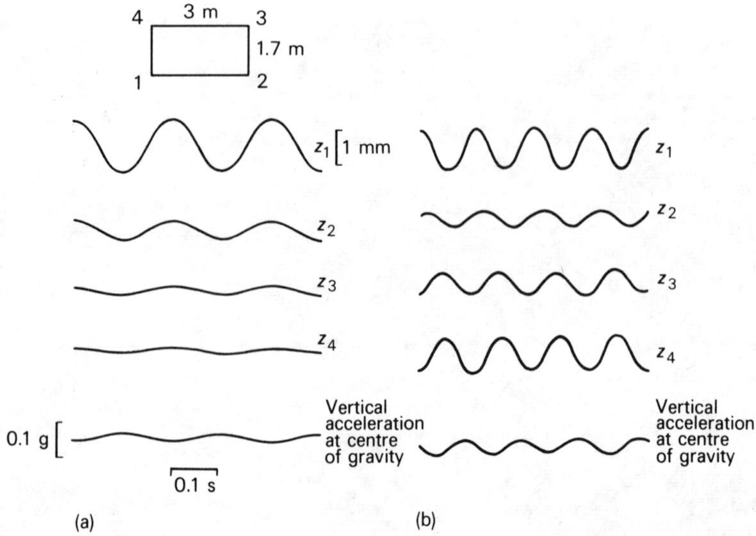

Fig. 5.14 Responses for the 500 kg flexible chassis vehicle: (a) low frequency decoupling of the four-magnet control loops for disturbance input in corner number 1; (b) cross coupling at near natural frequency for sinusoidal disturbance in corner number 1.

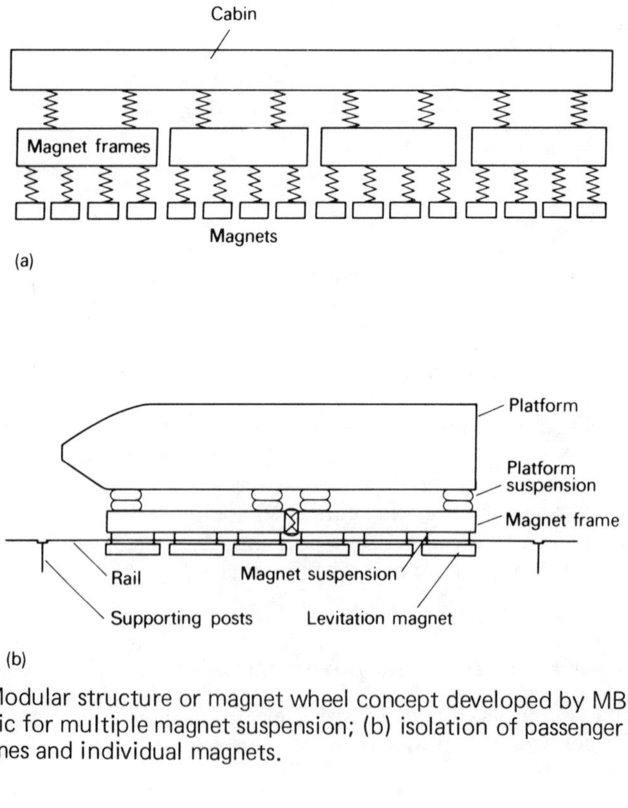

Fig. 5.15 Modular structure or magnet wheel concept developed by MBB: (a) schematic for multiple magnet suspension; (b) isolation of passenger cabin, magnet frames and individual magnets.

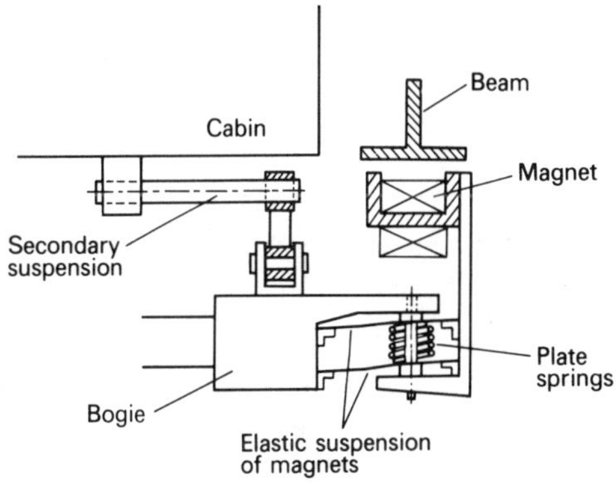

Fig. 5.16 Mechanical details of the magnet wheel concept on KOMET II suspension system (MBB).

transducers, magnets and power amplifiers, are subject to criteria of dynamic performance, as are components of any other control system. When the suspension system forms part of vehicle mounted equipment, it is then subject to a multitude of other criteria including weight, power consumption, reliability and ruggedness as well as safety requirements. Many of these requirements are mutually conflicting and therefore the process of optimisation is difficult. However, the amount of effort devoted to the development of vehicles during the 1970s was such that remarkable progress has been made in the component field and a review of some of these is likely to prove useful. This is particularly so because the applications of DC electromagnetic suspension range beyond that of passenger carrying vehicles to contactless bearings and support systems.

5.4.1 Transducers

In a magnetic suspension system transducers are required for the measurement of position, velocity and acceleration. In addition the flux control loop described in Section 5.2.2 also requires a transducer to measure the gap flux density. Whilst a wide variety of transducers appears to have been advocated, only two distinct types now seem to be in favour. These will be described later in this section. However all transducers need to be evaluated according to some criteria, and the following are some of the important ones. In setting out these criteria it is worth bearing in mind that in a control system, the transducers operate outside the feedback loop. The accuracy of the transducers cannot be improved by any compensation techniques and the control system is at best only as good as the transducers in it.

i) Bandwidth: this may be as large as 1000 rad s^{-1}.
ii) Robustness and stability under various operating conditions.
iii) Linearity over the operating range.
iv) Immunity from external noise, radiation and strong magnetic fields.
v) Ability to operate without mechanical contact.

Optical transducers

The simplest type of position transducer that can be used is the one illustrated in Fig. 5.5(a) or Fig. 5.6(a). This relies on the light falling on the phototransistor being intercepted by the suspended body. Whilst a collimated beam is desirable for improved performance or accuracy of positioning, it is not impossible to obtain suspension without these refinements. A further refinement which can be made where ambient light levels are high and variable is to use an infrared light source. With this transducer the output is linear over a 2 mm range of motion but the bandwidth is large, up to 10 MHz. Fig. 5.6(b) shows the circuit incorporating an optical transducer, and because the output is DC, lead compensation (differentiation) can be applied without much difficulty. If a filament type bulb is used for the light source, running the bulb at a voltage slightly lower than its rating increases the lifespan considerably. However, the transducer is unsuitable where dust or dirt might accumulate on it or where the suspended body is required to move and thus the beam of light liable to be interrupted.

Inductive transducers
Several alternative transducers have been constructed using the basic feature that the inductance of an iron cored (or air cored) coil is dependent on its proximity to a ferromagnetic body, or more particularly on the airgap between the two. The transducer measuring element itself is usually an E-shaped or U-shaped ferrite core or high grade lamination steel, such as mumetal with several hundred turns on it. In the system shown in Fig. 5.17 the inductance coil forms part of a Wien Bridge oscillator. The inductance is tuned with a suitable capacitance and as the airgap varies, so the frequency of the oscillator changes resulting in a variation in output with position. The bandwidth of the transducer is liable to be relatively low.

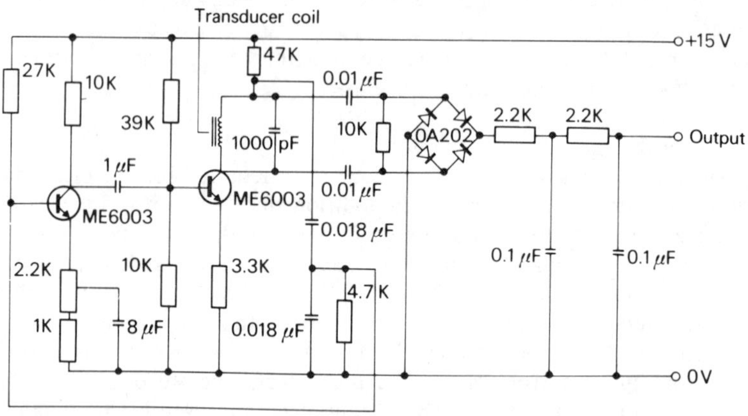

Fig. 5.17 Circuit for variable frequency tuned circuit position transducer.

For an oscillator frequency of 2 kHz, a bandwidth of only 20 Hz has been quoted.[169] A variant of this scheme is to use a separate oscillator, in which case the transducer with the tuning capacitor is placed in the feedback path of an operational amplifier. The circuit is shown in Fig. 5.18. Although the transducer is reasonably stable, immune from noise and linear over a wider range than the previous arrangement, like all tuned circuit systems containing iron cored inductances, it is susceptible to the ferro-resonant jump phenomenon.[127]

A more widely used form of inductive transducer[102,168] is constructed using

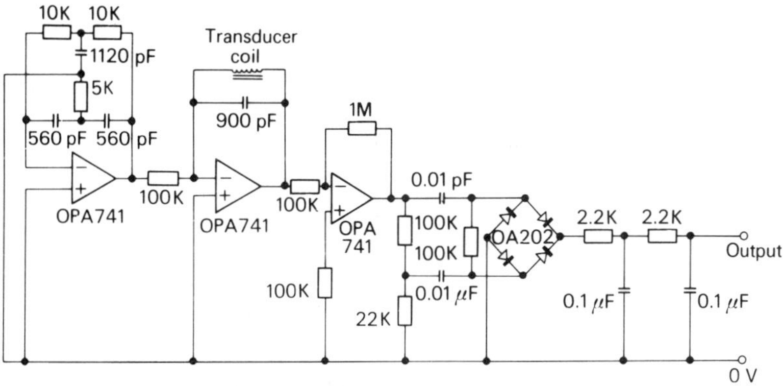

Fig. 5.18 Tuned circuit position transducer.

two identical coils, one being the transducer and the other one a reference coil (see Fig. 5.19). The coils are elements of a Maxwell bridge which is in balance when the inductance of the two coils is equal. Any imbalance of the bridge due to airgap variation of the coil used for the transducer is fed to a phase-sensitive detector to discriminate the direction of displacement. The bandwidth is dominated by the filter circuitry but the transducer is reasonably linear. However a rather serious drawback of this transducer was observed to be its dynamic response to a step such as that likely to be presented to it in the rail of a moving vehicle. Fig. 5.20 shows qualitatively the result of a test conducted on such a transducer. It can be seen that the transducer output is such as to overcompensate for the gap reduction.

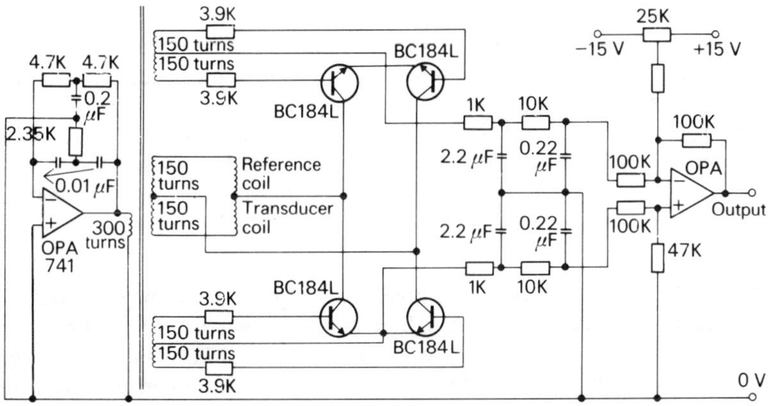

Fig. 5.19 Inductive position transducer with bridge and phase sensitive rectifier.

This overcompensation is due to the effect of the eddy currents induced in the rail being superimposed on the straightforward increase of inductance arising out of airgap reduction. The usefulness of such a transducer for the vehicle application must therefore be qualified.

Another variant of inductive transducers relies solely on the effect mentioned above. This is the eddy current proximity transducer, which is commercially

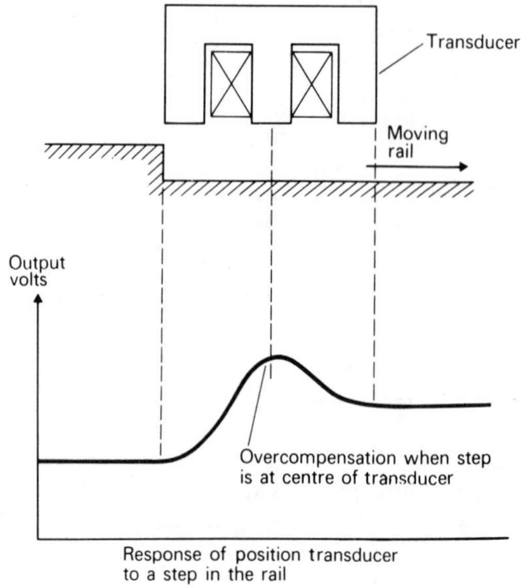

Fig. 5.20 Response of an inductive position transducer to a step in the rail.

available but, in magnetic suspension applications, appears to have been used only in the General Motors vehicle. The principal disadvantage is that a non-magnetic strip or rail is required in addition.

Capacitive transducers
Capacitive transducers offer the advantages that they are easily shielded from stray fields and also are not susceptible to changes of temperature. The transducer is essentially a single plate capacitor and therefore relies on changes of very small capacitance, and so very sensitive detection techniques are required. A technique using a nonlinear twin T-network is described by Lion[170] (see Fig. 5.21). In the transportation vehicle application, the return path would also prove to be quite a difficult problem.

Magnetic transducers
A magnetic transducer consists essentially of a permanent magnet with a Hall plate on its face, measuring the gap flux density when placed in the proximity of another ferromagnetic surface. The closer the magnet is to the rail, the greater the flux density. This is partly due to the change in the operating point on the demagnetisation curve of the magnet and partly due to the leakage flux being higher at larger gaps.

A circuit developed by Hodkinson[171] takes the Hall voltage V_H and uses it to control the Hall current I_H such that V_H remains constant. In effect the flux is multiplied by a current proportional to the output voltage to achieve linearisation. The essential outlines are shown in Fig. 5.22. In this figure the output of the amplifier A_1 is V_H ($= B \times I_H$, where B is the flux density and I_H is the Hall plate current), times the amplification factor of A_1. If R is the total resistance as seen by the amplifier A_2, we have

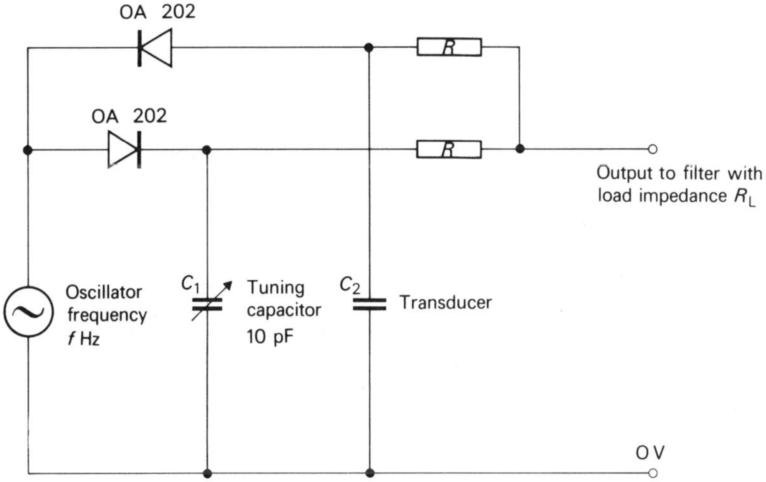

Fig. 5.21 Capacitive position transducer.

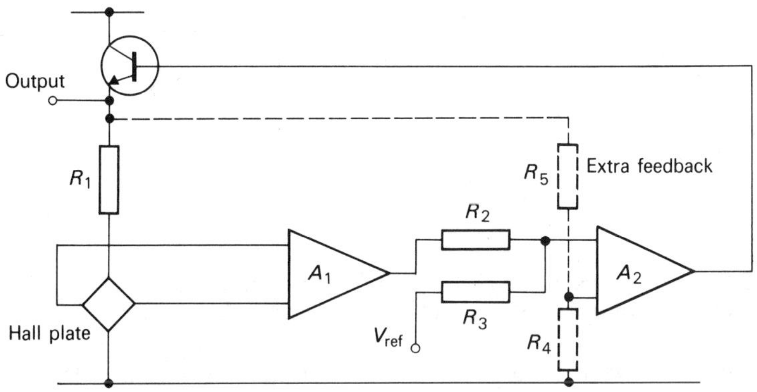

Fig. 5.22 Hall plate position transducer.

$$(V_H + V_{ref})A_2 = V_0$$

Therefore

$$\left(\frac{BV_0}{R} + V_{ref}\right)A_2 = V_0$$

Hence $$V_0\left(\frac{BA_2}{R} - 1\right) = -A_2 V_{ref} \tag{5.14}$$

When $A_2 \gg 1$ this approximates to

$$V_0 = -V_{ref}\frac{R}{B} = \frac{\text{constant}}{B}$$

When the gap is small, $B \propto 1/z$ and the output voltage is thus linearly proportional to the gap. The broken line connection shown on Fig. 5.22 is an alteration which consists of feeding a fraction α of the output of amplifier A_2 to its positive input terminals, and this modifies equation (5.14) to

$$V_0\left(\frac{BA_2}{R} - 1 - \alpha A_2\right) = -V_{\text{ref}}A_2 \tag{5.15}$$

i.e.

$$V_0 = \frac{V_{\text{ref}}}{(B/R) - \alpha} = \frac{C_1 z}{C_2 - C_3 z} \tag{5.16}$$

where C_1, C_2 and C_3 are constants. The relationship between L_{H}, the Hall plate current, and the gap z, without this modification, is of the form shown in Fig. 5.23(a), whereas the modification results in the curve shown in Fig. 5.23(b) and by an appropriate selection of the constants C_2 and C_3, a virtually linear output can be obtained for a range of $z < C_2/C_3$.

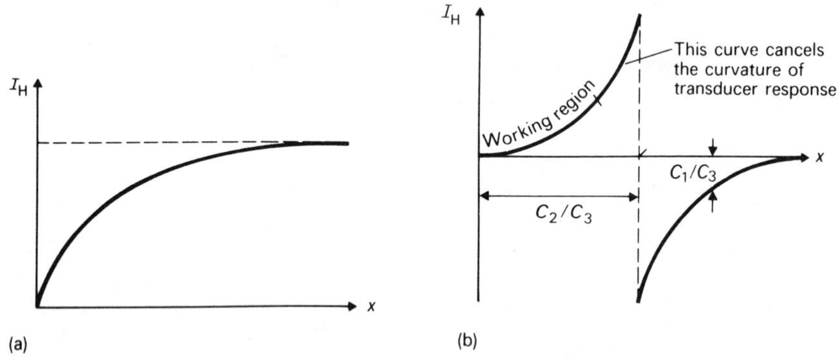

Fig. 5.23 Characteristics of the Hall plate position transducer: (a) Hall plate current with gap; (b) effect of extra feedback shown in Fig. 5.22.

The principal advantages of this transducer are that the output is DC, it appears to be virtually immune to noise and is of fairly rugged construction, but its performance is temperature dependent. Depending on the exact configuration of components around the amplifier A_2, in particular, the bandwidth varies between 1 kHz and 3 kHz.

A variant of the magnetic transducer developed by Whorlow[172] from the one described above has been called the I/B transducer. This has been used successfully in instrument systems, eliminating the need for a separate permanent magnet. Consider the arrangement of components as shown in Fig. 5.24, where the Hall plate is mounted on the face of the lift magnet. Obviously the Hall plate current I_{H} in this case is given by

$$I_{\text{H}} = V_0/R \tag{5.17}$$

As before the Hall plate voltage is given by

$$V_{\text{H}} \propto I_{\text{H}}B \tag{5.18}$$

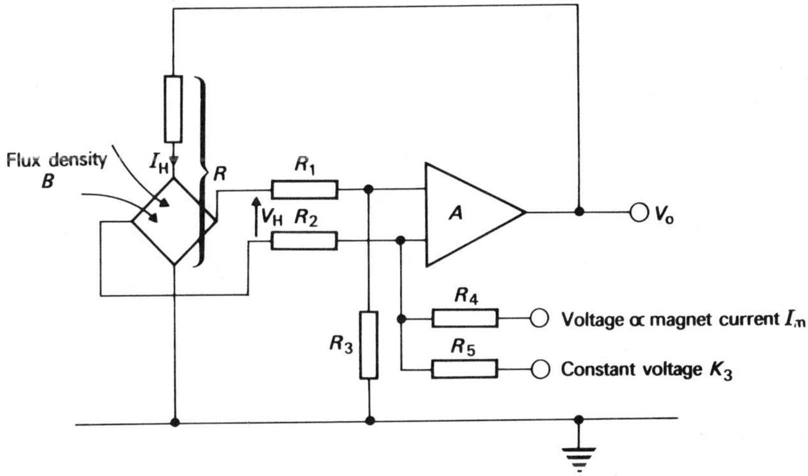

Fig. 5.24 The *I/B* transducer.

Also $\quad V_0 = -A(k_1 V_H - k_2 I_m - k_3) \qquad (5.19)$

Substituting equations (5.17) and (5.18) into equation (5.19) gives

$$V_0 = -A\left(k_1' \frac{BV_0}{R} - k_2 I_m - k_3\right)$$

Therefore

$$V_0\left(1 + Ak_1' \frac{B}{R}\right) = A(k_2 I_m + k_3)$$

When the amplification factor A is large,

$$V_0 = \frac{R}{k_1' B}(k_2 I_m + k_3)$$

$$= k_A\left(\frac{I_m + k_B}{B}\right) \qquad (5.20)$$

If $k_B = 0$ then $V_0 = k_A I_m/B$. Thus the transducer measures the quantity I/B, where I is the magnetic current and B the gap flux density it produces.

Now the ampere turns are given by

$$NI = \text{flux} \times \text{reluctance}$$

$$= B \times A \times S$$

where A is the pole face area and S the reluctance of the airgap, which is twice the suspension gap for a U-magnet. Thus

$$S = NI/AB$$

However,

$$S = 2z/\mu_0 A$$

where z is the suspension gap so

$$z = \mu_0 NI/2B$$

Thus for a given magnet the I/B quantity measures the gap, and in the circuit of Fig. 5.24, the Hall plate also acts as the divider.

The effect of the constants k_2 and k_3 may be examined by reference to the equation

$$V_0 = \frac{R}{k_1' B}(k_2 I_m + k_3)$$

When k_2 is zero,

$$V_0 = Rk_3/k_1'B = \text{constant}$$

Hence B is a constant independent of the suspension gap, and the system acts as a flux loop.

When $k_3 = 0$,

$$V_0 = \frac{Rk_2 I_m}{k_1' B} = \text{constant}$$

and the system acts as a position loop.

One of the disadvantages of the I/B transducer is that lead compensation applied to the output signal from the transducer does not give a true velocity signal[172] and thus it is necessary to provide a separate velocity transducer, particularly for systems requiring more stiffness than that possible through increased k_2. There is a limit on how large k_2 may be before the onset of instability.

5.4.2 Magnets

The design of magnets, particularly for providing lift in magnetic suspension systems, involves consideration of various interacting parameters. The design process may be divided into two stages. In the first stage, consideration is given to factors such as methods of generating magnetomotive force, suitable coil materials, the closed loop performance and the overall system requirements. There are then other factors which affect the detailed design. Of particular concern are the lift-to-power ratio, lift-to-magnet weight ratio, flux saturation effects, power losses and magnet weight in the case of application to vehicles.

Static design considerations

The starting point of the design procedure is to consider the equation giving the force of attraction

$$F \propto (NI)^2 \mu_0 A/z^2 \tag{5.21}$$

where NI = ampere turns; z = (suspension) airgap; A = pole face area. The constant of proportionality is determined by the detailed geometry of the magnet and track configuration.

Additional factors must then be considered such as saturation in the magnet core, flux leakage, power loss in the coils and due to eddy currents, and power input per unit lift capability. Changes in one parameter result in changes in the operating conditions and a fairly good way of illustrating these interactions[169] is shown in Fig. 5.25. This diagram illustrates changes which the designer can make as distinct from those which are consequential upon those changes.

The useful force available from a lift magnet depends upon the amount of flux which links the magnet and the reaction surface (rail). Equation (5.21) holds under idealised conditions of infinite permeability of the iron paths and thus absence of any leakage flux. In practice, the iron used for magnets and rails is not even high permeability special iron, but ordinary mild steel. The iron paths therefore possess reluctance of significant magnitude compared to the airgap reluctance. This in turn leads to substantial leakage fluxes and results in flux densities in the limbs of the magnet considerably higher than the airgap flux density. A typical figure may be up to three times higher and thus core saturation is bound to occur, especially where the magnet weight is also a constraint, as in vehicle application. Another aspect of magnets for vehicle application is that there will generally be a limit on the width of the track and thus on the overall width of the magnets. If U-shaped magnets are used (see Fig. 5.26(a)) there is a limit on the width of the pole faces at A. Excessively wide pole faces reduce the window area but more significantly increase the leakage flux between the two limbs, leading to a reduction in the airgap flux density. This effect is more accentuated as the length of the magnet is increased; the effect is illustrated in Fig. 5.26(c).

A considerable improvement in magnet performance can obviously be obtained by using better quality magnetic materials, but even for ordinary mild steels, saturation levels can be increased by as much as 10% by annealing the magnet core after matching. Another technique of enhancing the performance of the magnets is to use a stepped core as shown in Figs 5.26(b). Ideally the limbs should be tapered. It can also be seen from Figs 5.26(a) and (b) that there are two ways of locating the coils and the disadvantage of the first method, i.e. increased core flux density, may also be noted.

As in any electromagnetic device it is important to consider how the magnet is constituted in its balance between the magnetic materials and the electrical materials. The choice of the electrical materials devolves between copper or aluminium for the current carrying conductors. The choice of aluminium, particularly in the foil form, with anodised oxides as the only interturn insulation is very attractive. This gives enormously improved thermal properties, because the coils are almost solid blocks, and better utilisation of the area available for conductors, but results in essentially low voltage, high current magnet designs. In the transport application, provision of low voltage DC is not an economically viable proposition. On the other hand reliability and very high voltage DC operation are not compatible factors in DC amplifiers using solid state devices, particularly power transistors.

The problem of cooling of the coil windings for magnets is not serious provided that flux densities of the order of 0.5 tesla in the airgap are considered. In some designs flux densities as high as 0.7 tesla or even 1 tesla were proposed, but these will almost certainly require forced cooling. It would not, on the face of it, appear to be a necessary or desirable complication to add to the system.

It may be seen from Fig. 5.25 that almost every parameter change affects the

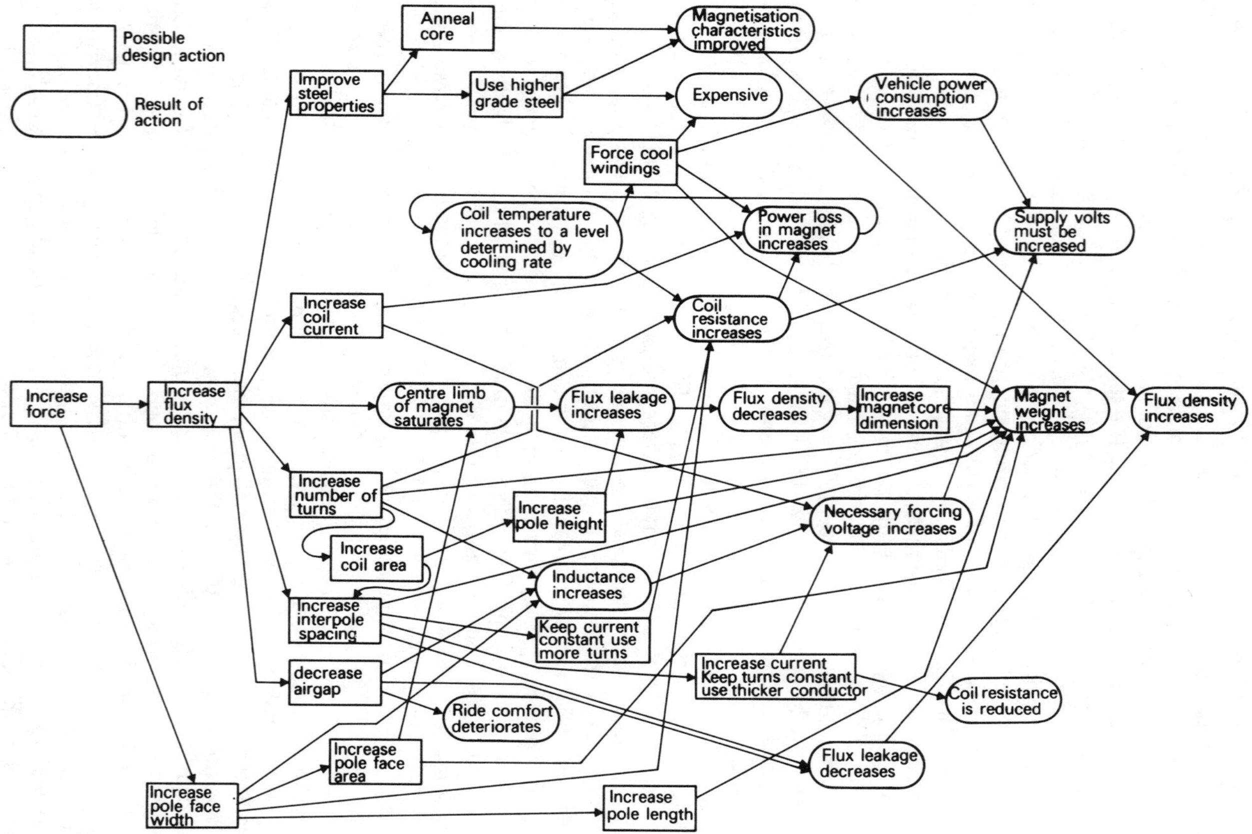

Fig. 5.25 Interrelation of magnet parameters.

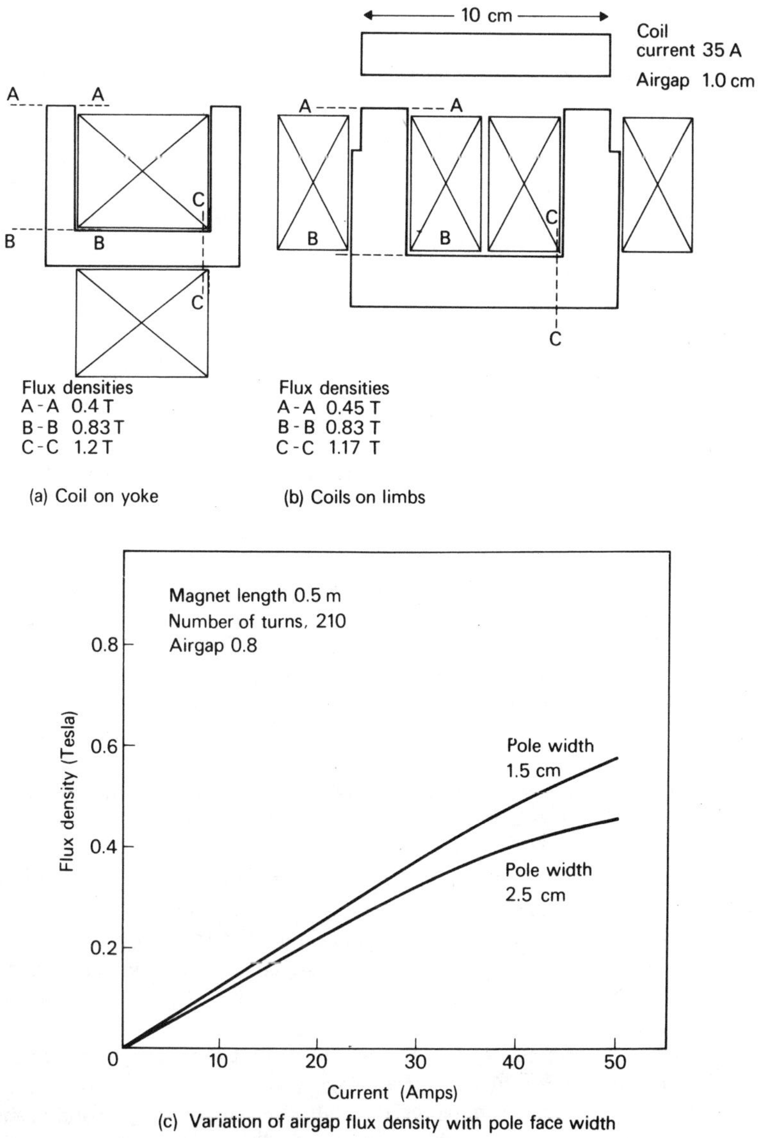

Fig. 5.26 Flux densities in U-shaped magnets in air and in iron.

weight of the magnet, therefore when changes are made to magnet designs their influence on the magnet weight must be borne in mind. The discussion so far has implicitly assumed that the magnet geometry is essentially that of Fig. 5.26, with the direction of travel of the magnet (vehicle) into, or out of, the plane of the paper. A substantial reduction in the weight of the magnet is possible for the same lift capability if the rail extends from left to right, and thus the magnet travel is at right angles to that implied in Fig. 5.26(a) and (b). However, for reasons explained in

Section 5.4.2 on p. 104, this advantage may be only illusory. For this direction of travel, the effects of induced eddy currents in the rail are such as to attenuate the airgap flux density substantially, leading to a serious loss of lift at high speeds.

Dynamic considerations

Magnets used for suspension are a part of closed loop control systems and it is essential to consider their performance in this context. Some of the factors discussed in static design also influence the dynamic characteristics. The following features need particular consideration:

(a) inductance of coils and the time constant;
(b) operating voltage and forcing ratio;
(c) lateral guidance forces;
(d) effect of eddy currents induced by relative movement of the suspended member with respect to the reaction surface;
(e) route switching of vehicles.

Although point (e) is essentially a dynamic feature, it is too involved with the geometry of the magnets and is dealt with separately. Ride control and ride comfort are also dynamic features of the control system as a whole, and these features are also dealt with separately.

Given a particular lift (load) requirement and a notional limit on the airgap flux density, the inductance is virtually beyond the control of the designer. For a given conductor material of resistivity ρ, it is a straightforward exercise to show that the time constant is determined by the 'window' area and that the ratio between the turns and current can have only second-order influence. This influence is not very quantifiable without detailed studies of the flux patterns such as those illustrated in Fig. 5.27, which of course have been computed for purposes of determining the forces generated by lateral displacement of the magnet with respect to the rail.[173] The only significant factor which influences the time constant is therefore the resistance of the coils. A typical value of the time constant for one magnet of the 1 tonne University of Sussex model, which has aluminium foil around 1 m long magnets and an airgap of 1.0 cm, is about 400 ms. The magnets used for vehicles are not likely to be very large. The tendency appears to be to use a large number of magnets of not much more than 1 to 2 tonne lift capability, and the time constants encountered are likely to be of similar orders of magnitude to the one quoted here.

The large magnet time constants involved mean that obtaining bandwidths of at least 10 Hz and higher for heavy suspensions presents quite a formidable task. Such bandwidths are only obtainable through the capability of forcing current variations, i.e. sufficiently large voltage forcing. This is dealt with again in the amplifier design considerations. Practical experience so far suggests that a supply voltage of at least three times the IR drop across the magnet is about the minimum, but five times is desirable.

An approximate notion of the behaviour of the magnet in relation to a given supply voltage may then be obtained as follows.[174] Rewriting slightly equation (5.21) gives

$$\text{Force } F = (NI)^2/\mu_0 A S^2 \qquad (5.22)$$

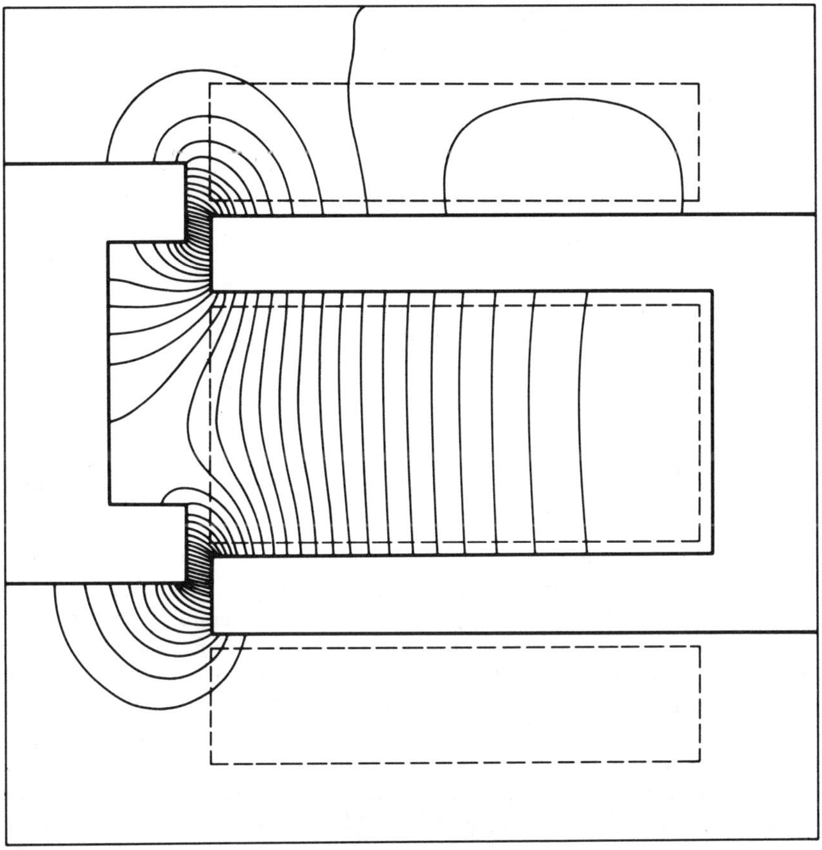

Fig. 5.27 Computed flux pattern for a U-shaped magnet against U-shaped rail.

where I is the quiescent current, A the pole face area and S the reluctance of the airgap, assuming infinite permeability for iron.

Therefore

$$\frac{dF}{dt} = \frac{2N^2 I}{\mu_0 A S^2} \frac{dI}{dt} \tag{5.23a}$$

The supply voltage is given by

$$V = RI + L\frac{dI}{dt}$$

or

$$\frac{dI}{dt} = \frac{V - RI}{L} \tag{5.23b}$$

From equations (5.23), therefore,

$$\frac{dF}{dt} = \frac{2N^2}{\mu_0 AS^2}\frac{I(V-RI)}{L}$$

Since $L = N^2/S$, we have

$$\frac{dF}{dt} = \frac{2}{\mu_0 AS}I(V-RI)$$

$$= \frac{I}{z}(V-RI) \qquad (5.24)$$

where z is the airgap.

If the voltage forcing ratio is $V_F = V/IR$,

$$\frac{dF}{dt} = \frac{I^2R}{z}(V_F - 1)$$

$$= \frac{P}{z}(V_F - 1) \qquad (5.25)$$

Equation (5.25) represents the slewing rate of the force generated by the magnet–amplifier combination. It appears on the basis of experiments that a factor of something like 25 times (per second) the static force generated by the magnet is required to give satisfactory response to force disturbances to the system.

The guidance forces generated by lateral displacement of magnets as shown in Fig. 5.27 are unlikely to be of a sufficient magnitude even if the airgap is kept constant. The problem is that as the magnet is displaced the effective lift area diminishes and in the vehicle application this will cause a reduction in the lift force and a corresponding increase in the airgap. The suspension control system makes correction for this and the variation of lift force with sideways displacement of magnet as well as that of guidance force is illustrated in Fig. 5.28. This figure also illustrates the point that if the US Department of Transportation figure of 42% of lift force, which has been arrived at after a great deal of experimental and theoretical work, is to be satisfied, then the sideways force may be inadequate. A scheme using lift magnets slightly displaced in the manner of Fig. 5.27, first on one side and then the other, thus forming a zig-zag line on each side, has been tested in both Germany and Japan to provide the velocity damping required in what would otherwise be an oscillatory system. Using the lift magnets to provide both lift and guidance forces clearly introduces cross coupling between the control systems of the two models. Whilst this might be adequate for the low speed urban transport system vehicles, there must remain considerable doubt as to the practicability of this scheme in high speed (exceeding 300 km h^{-1}) vehicles. For such vehicles, it is believed that separate guidance magnets acting on rails along the sides of the lift rails will be required. Most of the German vehicles produced to date have guidance magnets.

The fourth dynamic consideration mentioned at the beginning of this section is that of eddy currents induced in the rails by the movement of magnets with respect to the rails. This has a twofold effect. The first is obviously that it will

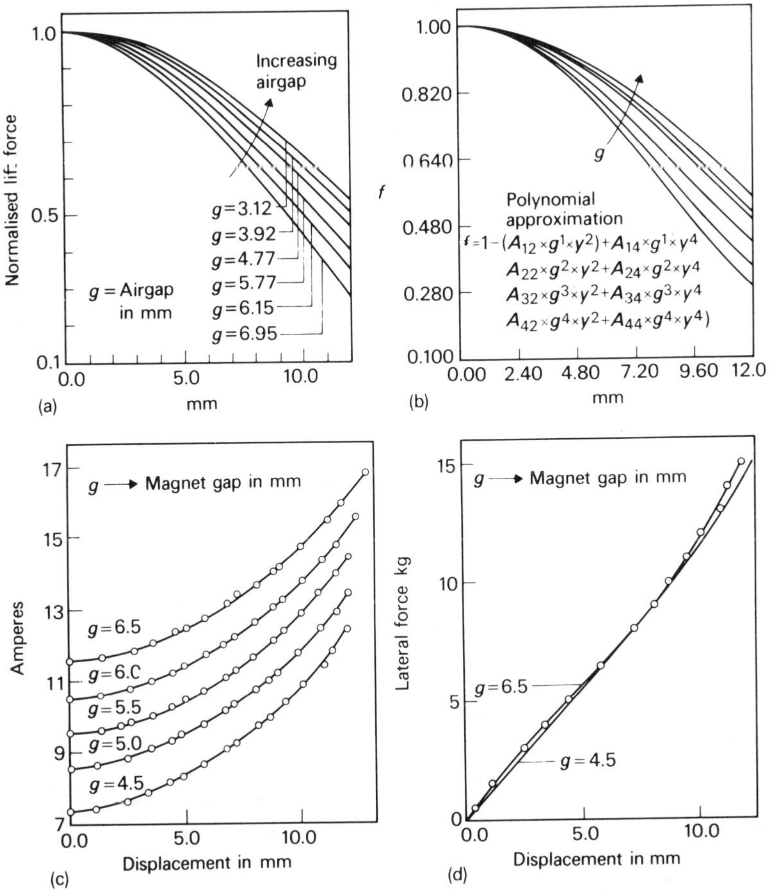

Fig. 5.28 Lift and lateral forces in U-shaped magnets: (a) variation in lift force with lateral displacement; (b) approximation to experimental data in (a); (c) variation of magnet excitation current with lateral displacement to keep lift force constant at different gaps; (d) lateral force as a function of lateral displacement at constant lift force.

produce a drag and thus require additional power to be supplied to the prime mover (linear motor) in order to overcome the drag. This drag may not be so serious as in the superconducting levitation systems described in Chapter 3. However, a more serious effect is that the eddy current reaction results in a reduction of the airgap flux density. Since the lift force is proportional to the square of the flux density B, there is an even greater loss of lift arising from the induced eddy current reaction.

There appears to be little published literature[175] on the subject of the eddy current problem in attraction suspension systems, partly in the belief that laminated rails will have to be used for high speed vehicles, and partly because the problem

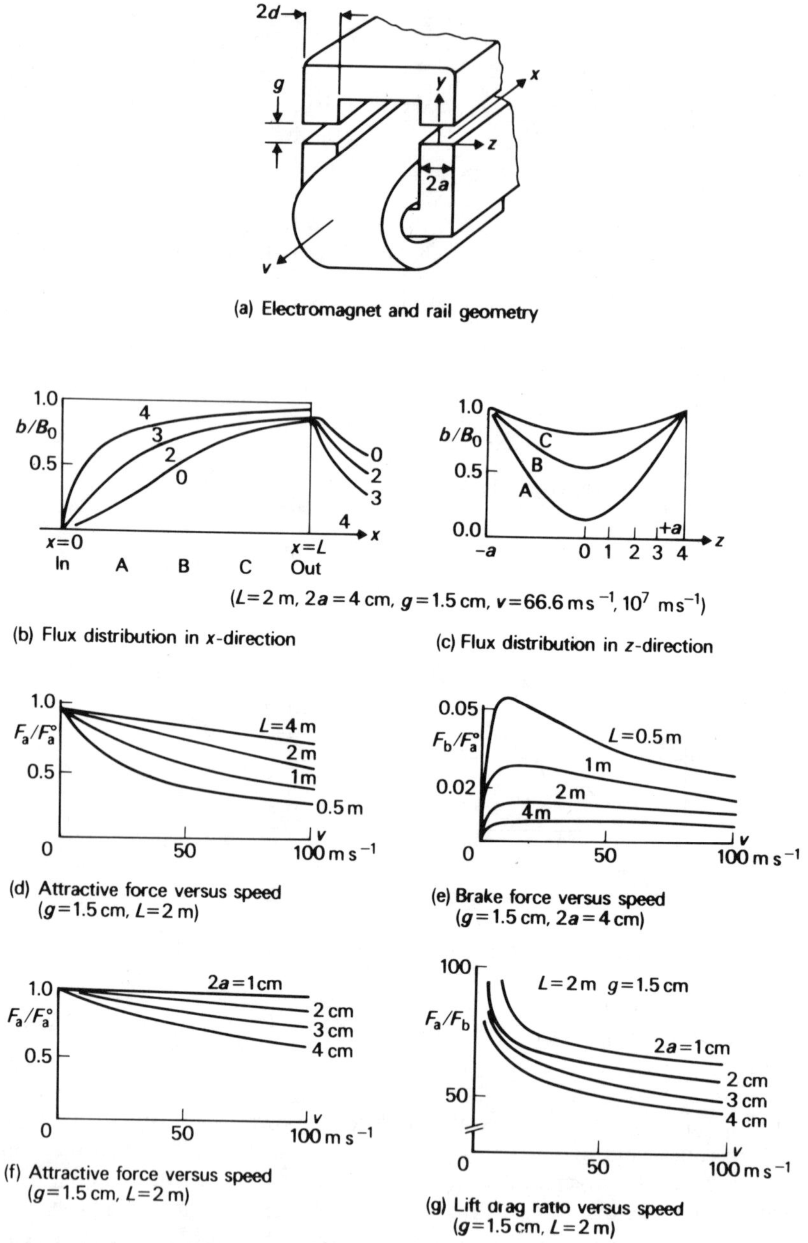

Fig. 5.29 Characteristics of U-shaped lift magnets.

is a genuinely difficult one to compute as well as evaluate experimentally. The computational difficulty arises from the fact that it is in essence a three-dimensional one. Some of Yamamura's results are reproduced in Fig. 5.29. These are in broad agreement with the results obtained by the author himself[176] in an experiment where a long steel rail was propelled past stationary magnets of various geometries.

The curves shown in Fig. 5.28 are self-explanatory but the implications of these in high speed ground transportation may not be. An observation to be made is that the longer the magnets the less are the adverse effects of induced eddy currents, i.e. loss of lift and drag. Two conclusions can therefore be drawn from these observations. Since long magnets are impracticable and a large number of closely spaced small magnets are known to behave at high speeds as one long magnet, because the rail remains magnetised, the vehicle construction is essentially that of using 1 to 2 tonne magnets even for vehicles of 30 tonne and over. The second conclusion from this observation is that a U- or E-magnet travelling in a direction such that the rail is subjected to alternate N–S polarity would make the problem of eddy currents considerably worse. Therefore the direction of travel of the magnet implied in Figs 5.28 and 5.26 is the only practicable one for high speed vehicles. A further disadvantage of the direction of travel at right angles to that implied in Fig. 5.28 is that the rail cross section has to be nearly the same as the pole face area of the magnet. The amount of material in the rail and the cost of the track are therefore liable to increase proportionately. If, in addition, since this (square) cross section has to be laminated in order to alternate the eddy currents, this places very severe limitations on the system. Another observation that needs to be made is that, provided a number of closely spaced magnets are used, it may not be necessary to use a laminated rail of U-cross section, with the attendant advantage that the drag force peak will be reduced. This in turn will mean that the rating of the propulsion unit will not be dominated by this drag peak as pointed out in the superconducting levitation systems. Finally, the increase of current required to compensate for the loss of lift will result in an increase of the power-to-lift ratio. Again there are no reliable guides as to how big this increase will be, but it is possible that it might be a factor of two or more.

Magnet configurations for route switching

The generally accepted configuration for the track is a T-shaped track and U-magnets as shown in Fig. 5.30. It will be observed that at a junction (switch) in the track, the supports for the magnets coming down from the vehicle body, which are quite substantial, will need to go through appropriate slots in the track. This will produce the situation as shown in Fig. 5.31(a). Where the radius is large the gap shown at the centre right could be of several vehicle lengths. A proposal by Krauss-Maffei in Germany was to provide the vehicle with double the number of magnets, i.e. eight magnets for a four magnet vehicle, and to place them in pairs alongside. The switch itself would have dual rails as shown in Fig. 5.31(b). There are two problems with this proposed solution. The first one is that the vehicle is required to carry an additional set of magnets and amplifiers, and the weight penalty might make this untenable. The second problem is more serious. In transition of the vehicle from the straight to the branch line, when the magnets are over that part of the track where previously there was a large gap, it is necessary (*i*) to switch the normal magnets off and the auxiliary magnets on and (*ii*) to switch the auxiliary magnets off and the normal magnets on. In operation (*ii*) it is insufficient only to switch the main magnets on again. The switching routine will

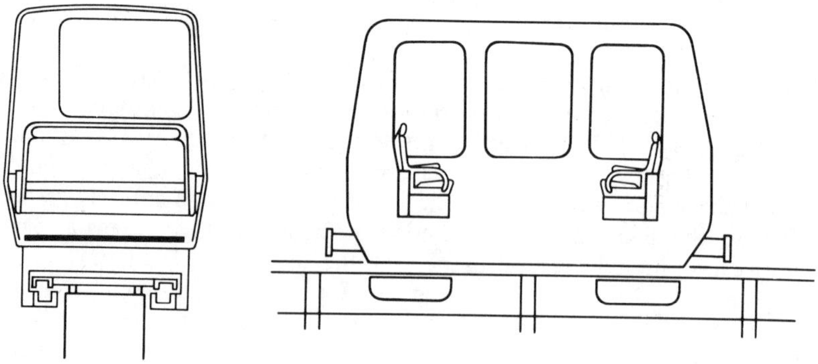

(a) (b)

Fig. 5.30 Disposition of magnets and magnet supports for vehicle with magnets underneath the floor: (a) end sectioned view showing U-shaped magnets and T-shaped track; (b) side view.

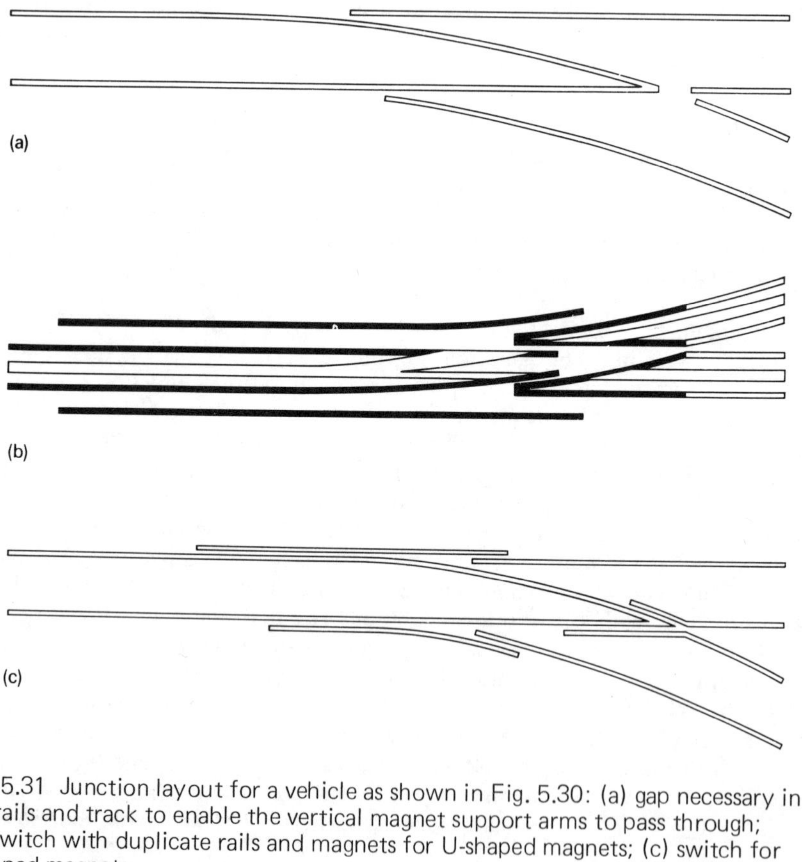

(a) (b) (c)

Fig. 5.31 Junction layout for a vehicle as shown in Fig. 5.30: (a) gap necessary in the rails and track to enable the vertical magnet support arms to pass through; (b) switch with duplicate rails and magnets for U-shaped magnets; (c) switch for I-shaped magnets.

ail in case the auxiliary magnets are not fully turned off. This scheme is therefore nlikely to be considered safe enough for a passenger carrying vehicle system.

The second method of achieving route switching is that proposed by the Germans for the Emsland track. This involves physically moving the track in much he same way as in a railway system. The third method, proposed by the author,[177] s illustrated in Figs 5.32 and 5.31(c) and uses novel I-shaped magnets. These magnets have been tested on a small scale prototype model (see Fig. 5.33), and hey will operate satisfactorily with an inverted L-shaped rail on one side or on the other, or on both. A careful study of Fig. 5.31(c) reveals that with the layout of duplicate rails shown, even in the critical portion of the track for U-magnets,

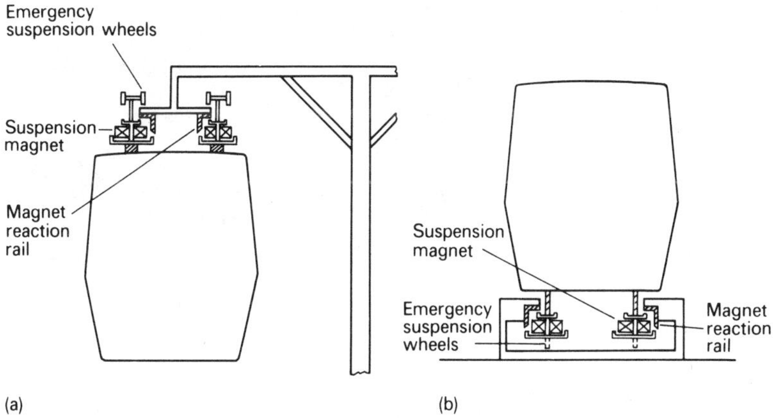

Fig. 5.32 Schematic of suspended vehicles using I-magnets: (a) overhead support; (b) underfloor support.

the I-magnet is always under rails. At this critical point the worst that can happen is that one half of the magnet may be under the rail on one side and the other half on the other. Tests have shown that under this worst condition, there is no impairment of lift capability of the magnet. There are two disadvantages of this geometry of the magnet. The first is that under normal operating conditions, i.e. when there is rail on one side of the magnet only, approximately half the magnetic circuit of the magnet is not utilised, constituting a dead weight. The magnet is likely to be 50% heavier than its corresponding U-magnet. Secondly an inverted L-shaped rail is not easy to fabricate. However, the side face of the rail can be used for separate guidance magnets. During operation it is envisaged that the direction of the vehicle(s) will be controlled at the switch by trackside magnets exerting a pull in the appropriate direction. The vehicle shown in Fig. 5.33 appears to be the only known example, admittedly in a model form, of an operating vehicle using any form of suspension or levitation and capable of negotiating a junction.

5.4.3 Power amplifiers

The recent rapid advances in the technology of attraction magnetic suspension are attributable almost entirely to the development of solid state electronic devices,

Fig. 5.33 An experimental 0.25 tonne vehicle with I-magnets on its track, showing the switch.

particularly power devices. High power-to-weight ratio and efficiency, coupled with extreme reliability of operation, are the principal requirements for magnetic suspension in almost any application: passenger carrying vehicles or any other described in Section 5.6. Therefore whilst the early experimenters[76,77] used valve amplifiers to suspend relatively large weights, it is not surprising that at the time this work did not lead to many (if any) practical applications.

It has been pointed out that the time constants of typical lift magnets may be as large as several hundred milliseconds, and yet the magnet–amplifier combination must act as a closed loop control system with a bandwidth of at least 10 kHz. A fairly substantial reserve voltage to force rapid current change in the magnet in order to overcome the inductive voltage is therefore an essential feature in the design of DC power amplifiers for controlled DC electromagnets. This requirement in turn can lead to large power dissipation and low efficiencies in quiescent operating conditions.

The choice of amplifier configuration is essentially between class A and class D. These two configurations are illustrated in Figs 5.34(a) and (b). The active linear regulator or the active switch is either a transistor or a thyristor. The class A amplifier has a smooth current regulation capability and is therefore obviously preferable to the class D where the on-off sequence is bound to interfere with the transducers, particularly the flux loop transducer which is mounted on the lift magnet face. However, when the efficiencies of the two schemes are considered it is quite clear which of the two options is the only practicable one. The following exercise for one particular magnet is instructive.

Assume that a lift magnet is required to lift a notional weight at 1.0 cm airgap

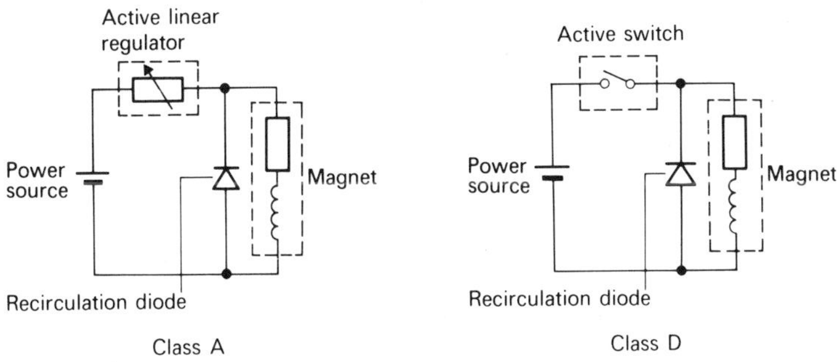

Fig. 5.34 Class A and Class D amplifier configurations.

and the current at this gap is 50 A. If the resistance of the magnet winding is 0.1 Ω, the steady state voltage drop across the windings will be 5 V. There are now two requirements to be considered. The first is that for the initial lift-off the gap will be 2.0 cm and at the very least 100 A will be required from the amplifier. The second requirement is that the forcing voltage necessary is, say, at least three times the steady state voltage drop. These two requirements mean that the HT voltage of the amplifier is 15 V of which, in the quiescent state, a 5 V drop is across the magnet but 10 V is across the power output transistors. The power dissipated in the magnet is $50^2 \times 0.1 = 250$ W, whereas the power dissipated in the amplifier is $50 \times 10 = 500$ W. The efficiency for a class A amplifier is therefore $(250/750) \times 100 = 33.3\%$.

The class D operation relies on the transistors or thyristors being turned fully on or off for variable periods (i.e. variable mark-space) and thus the average power is controlled. Let us assume that since the voltage is sufficiently low, transistors can be used whose turn-on time is 1 μs and whose turn-off time is 9 μs. It is also a reasonable assumption to say that during switching half the power is dissipated in the transistors. If the switching frequency is 5 kHz, the total switching time is $5000(9 + 1)$ μs = 50 ms, i.e. 1/40th of 50 A × 15 V is dissipated in the transistors, which equals $750/40 \simeq 20$ W. A further power loss to be accounted for arises from the resistive drop due to $V_{CE(sat)}$ of the switches and the forward voltage drop V_f of the recirculation diodes. This may be 1 V at 50 A, i.e. 50 W. The total losses are therefore 70 W. The power output is still 250 W and hence the efficiency this time equals $(250/320) \times 100 \simeq 80\%$. The average power supply current also is now reduced and equals $320/15 = 21.3$ A. The class D configuration, therefore, clearly offers advantages in terms of power dissipation in the amplifier, efficiency and even the rating of the power supply.

On the other hand there are some problems and disadvantages intrinsic to the class D configuration. Firstly there is the increased complexity of the pulse width modulator, although it offers the advantage of isolation between control circuits and power circuits using either transformers or opto-isolators. Secondly, there are the problems of switching, such as audio noise in the magnet and some RF radiation which may interfere principally with the transducers and the control circuits.

The selection of the configuration is likely to be determined by several other constraints. For example, the DC HT supply is likely to be rectified 3-phase AC for any industrial or transportation application. Until recently the only devices ava able for operation at such a high DC voltage (500 V to 600 V) would have be thyristors. The situation is changing very rapidly and transistors of appropria voltage rating are coming on the market for lift magnet–amplifier combinatio of 1 to 2 tonne capability. There is also a limit on the voltage across the magn winding: the higher the voltage, the bigger will be the need for insulation, and increase of operating voltage soon becomes counterproductive. One of the ma other requirements that may be mentioned is that of forcing down the current the magnet. The larger the magnet to be controlled, the more inadequate will the provision of only the recirculatory diode connected across the magnet. In t positive direction the forcing ratio has been defined as $F =$ supply voltage/stead state voltage drop. This is a way of expressing the voltage available to overcom the inductive voltage drop. There are difficulties in defining negative forcing in similar way since the inductive voltage is now in a direction opposite to that of t resistive drop. Neglecting the voltage drop across the diode the maximum voltag available to force the current negatively is that which exists across the resistanc (see Fig. 5.35). The same comments apply to class D amplifiers. The diode in th case must be a high power one as it is used continuously at high frequenc

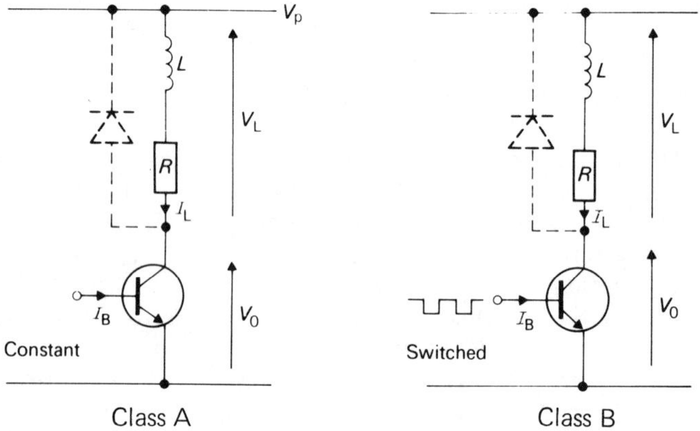

Fig. 5.35 Single transistor amplifiers with an inductive load.

A novel circuit (see Fig. 5.36) was used by Hodkinson[169] to provide negative forcing for a class D amplifier using the single supply voltage effectively in reverse. Both transistors are turned on together giving a load voltage of

$$V_L = V_p - 2V_{CE(sat)}$$

When the transistors are turned off the load current recirculates through the diodes D_1 and D_2 and the load voltage becomes

$$V_L = -(V_p + 2V_D)$$

The same circuit could also be used for class A amplifiers. The circuit has one disadvantage in that power is now dissipated in two transistors during saturation and hence the efficiency of the amplifier will be reduced.

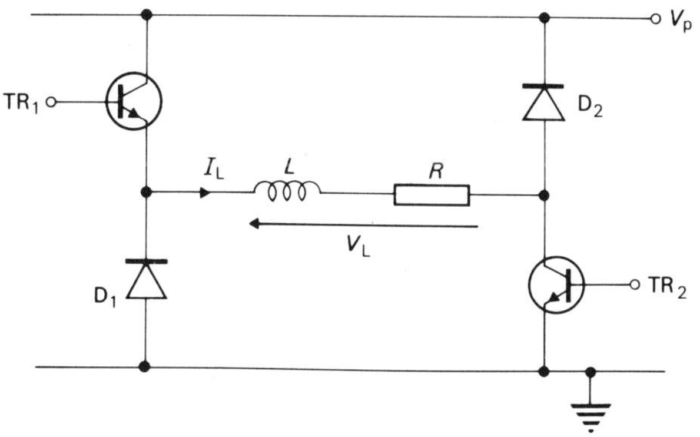

ig. 5.36 A Class D amplifier with positive and negative forcing capability.

.5 Ride control and passenger comfort in passenger carrying vehicles

'he problem of passenger comfort and ride control are immensely complicated ecause on the one hand they depend upon subjective feeling of well-being in a assenger cabin and on the other they depend on design parameters of considerable nportance, such as the suspension and guideway dynamics. An attempt is made ere only to present a broad picture of some of the parameters which have a ominant influence.

Ride quality and ride standard

'he ride is defined as a general feeling of well-being when travelling in a vehicle, nd amongst the factors affecting this, most important is the vertical acceleration. evels based on human tolerance limits are shown in Fig. 5.37. In the power spectral ensity (psd) specification the variation of track roughness from a nominal reference is assumed to be random. Such a representation is suitable for theoretical omputation of cabin acceleration levels. In the discrete frequency specification a inusoidal variation of track roughness is assumed and this is more suitable for xperimental verification. The various suspension parameters should be designed o that the effective vertical acceleration remains below the specified value in the ntire range of frequencies.

Suspension stiffness

The lift-to-suspension power ratio reduces with the increase of operating airgap or track clearance. Change in airgap per unit change in load, which may be defined as suspension stiffness, is therefore an important index for efficient operation of magnetic suspension systems. The stiffness coefficient is proportional to the steady state error[164] in the airgap, and can be increased either by higher feedback gains or by incorporating an integral of error action. There is an upper limit on the increase of feedback gain attainable due to noise and saturation effects at high gains: the latter method, whilst theoretically providing infinite stiffness, may result in a deterioration of the transient performance.

Operating speed and airgap

The choice of operating speed is primarily dependent on the desired system capacity

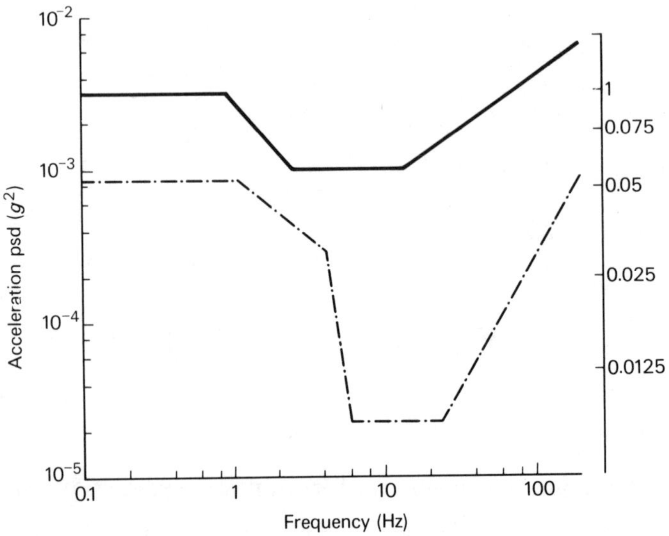

Fig. 5.37 Ride quality specifications: —·—·— based on random track roughness (power spectral density acceleration levels); ——— based on discrete frequency approach (rms acceleration levels).

and safety requirements. The longitudinal speed is also correlated to ride quality and mean operating airgap (see Fig. 5.38). Comparison of Figs 5.37 and 5.38 shows, for example, that the required airgap for urban transit vehicles over the roughest track is around 12 mm. This nominal value of airgap is, as seen in Section 5.4.2, an important parameter in the design of lift magnets.

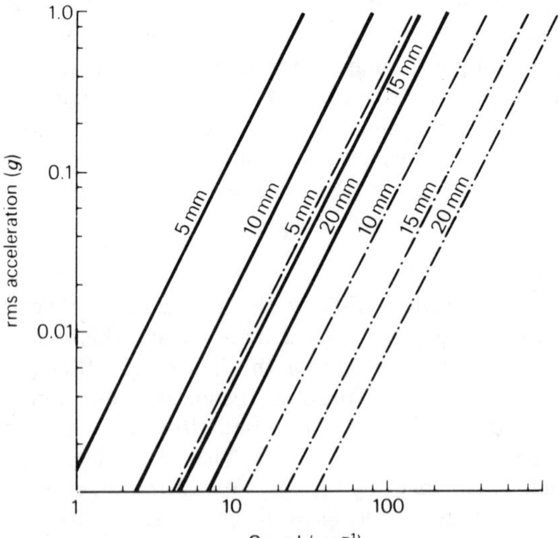

Fig. 5.38 Variations of peak acceleration with speed and mean operating airgap: —·—·— smoothest track; ——— roughest track.

Damping of the suspension system

Ride characteristics of any transportation system are closely related to the overall damping of the system. It has been estimated that if the heave dynamics can be described by a second-order differential equation,[164] the range of desired damping ratios for a DC electromagnetic suspension system is 0.5 to 0.8. Beyond this range the improvement to ride characteristics is marginal.

Natural frequency and mode control

For reasons of economy and aesthetic design it is assumed that the guideway might consist of simply-supported spans 30 m long. Such beams would introduce an additional mode into the overall suspension system (see Fig. 5.39(a)) resulting in a dynamical interaction between guideway vibration and suspension control system(s). From a practical point of view it is undesirable to design a system so that a small variation in excitation frequency brings about a resonance condition. This danger can be minimised by sufficient separation between two natural frequencies.[178] The degree of separation dependent on the mass ratio of the two subsystems (see Fig. 5.39(b)) will have to be attained through the design of the vehicle control system and the guideway. However, the choice of suspension

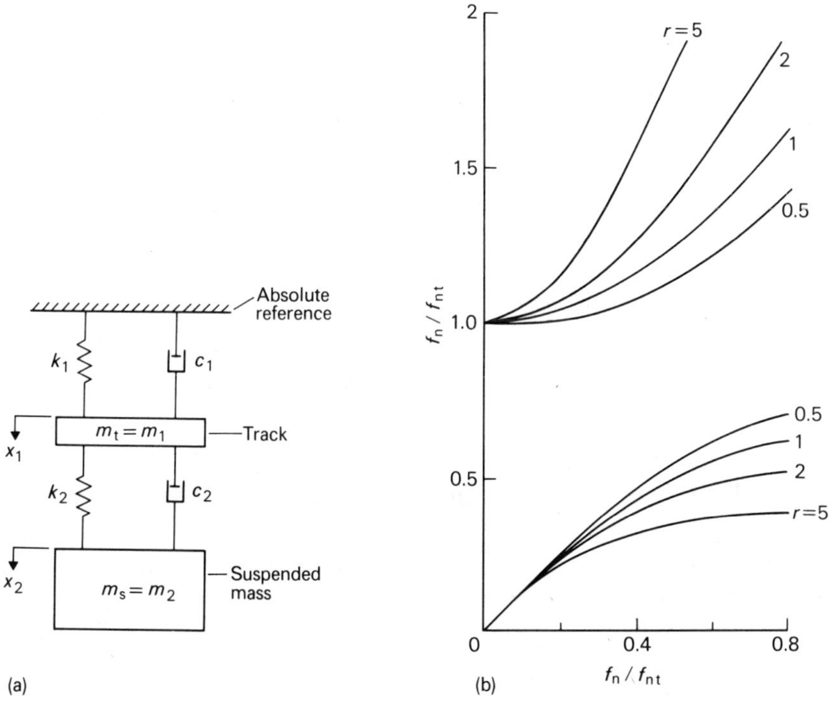

Fig. 5.39 Flexible guideway and closed loop suspension system: (a) lumped representation; (b) separation of natural frequencies of system in (a) for various mass ratios m_t/m_s.

natural frequency is related to the required stiffness and lift capability of the magnets, therefore the guideway design cannot be isolated from the design of the vehicle control systems. For the type of guideway referred to on p. 121, the natural frequency of the track $f_{nt} = 3$–5 Hz and for an appropriate separation, the natural frequency of the suspension system may be 6–10 Hz, as an example.

Guideway mass and damping

Since the stability of the overall system as shown in Fig. 5.39(a) is a fundamental requirement, it is necessary to establish an explicit relationship between guideway mass, damping and stability margin of the composite system. Analysis shows[162] that for one extreme case where the mass of the track $m_t \gg m_s$, where m_s is the mass of the suspended vehicle, the track and the suspension system become nearly decoupled. At the other extreme, $m_t \ll m_s$, the dynamics of the overall system are nearly independent of the characteristics of the track. When $m_t \simeq m_s$ the overall system is greatly affected by a small variation of the mass ratio. These observations are verified in experiments.

The coupling between track and suspension vibrations can also be controlled through the damping of the guideway. If the guideway damping is increased, damping of both low and high frequency modes is increased up to twice the critical value of the damping factor. Beyond this the damping of the low frequency (track) mode continues to increase whilst the damping of the high frequency mode (suspension) rapidly decreases.

Track irregularities

Economic considerations dictate that a wide tolerance on the alignment of different spans of guideway may be necessary. Such misalignments could represent 1 Hz square wave disturbances with up to a maximum of ± 20% of the mean operating airgap for the spans of 30 m when the vehicle is running at 70 km h^{-1}. Another track disturbance input which may be even more difficult to handle is a midspan deflection of up to 50% nominal airgap. Acceptability of any feedback control system will depend upon its capability of maintaining stable suspension whilst keeping the vertical acceleration level below the permitted value, at least in the passenger cabin. It would appear that at speeds in excess of 300 km h^{-1} both secondary suspension and a more rigid and closely aligned track will be necessary.

5.6 Application of magnetic suspension using controlled DC electromagnets to contactless bearings or support

At the time of writing the author's team is engaged on several industrial applications or investigations in areas other than passenger carrying vehicles. One of the early investigations was that of an application as an anemometer capable of measuring wind speeds of the order of 1.33 m s^{-1} (3 mph). A prototype of this instrument is shown in Fig. 5.40. It consists of the central element carrying the turbine rotor suspended between two open-ended pot core shaped controlled DC electromagnets. A special feature of this instrument is that it uses the I/B transducer described in Section 5.4.1 on p. 102 and thus has a Hall plate in the face of each magnet. Other applications investigated in the area of contactless bearings are a dumb-bell shaped rotor rotating at 55 000 rpm for synthetic fibre applications, a rotor for a

Fig. 5.40 Magnetically suspended anemometer using *I/B* transducer.

turbomolecular pump and an ultra-centrifuge. An application where rotation of the suspended object was precluded is illustrated in Fig. 5.41(a). The system consists of a suspended yarn package. Yarn from this bobbin is taken down to the bottom and up through the package. As shown in Fig. 5.41(b) there are also two fixed bobbins, and yarn from these is led through two fliers which rotate in opposite directions forming an intersecting figure-of-eight loop and also a balloon round the suspended yarn package. All the three strands are pulled through a hole in the suspension system and the object of the exercise is to form a novel kind of sewing cotton in which the three yarn strands are plaited. In conventional sewing cotton the individual strands are held together by pretwisting each strand, and all the strands become bound when they uncurl. The largest suspension system built by the author's team is the suspension of an 8 tonne drum which is also driven by a linear induction motor (see Fig. 5.42). There are two suspension magnets and four guidance magnets on the framework. The suspension magnets themselves weigh 2 tonne but the suspension power is, surprisingly, only 800 watts (for supporting 4 tonne) per magnet. A few other similar ventures are in the discussion stage at the moment.

5.7 Conclusions and future prospects

It is clear that controlled DC electromagnet suspension is at a more advanced state of development than any other suspension scheme proposed for passenger carrying vehicles. One of the drawbacks of this scheme is the small operating airgap, but the German development and perhaps the Japan Air Lines development might illustrate how serious this is likely to be. A second drawback of this scheme is the lack of a proven route switching scheme, except in a model form in the author's

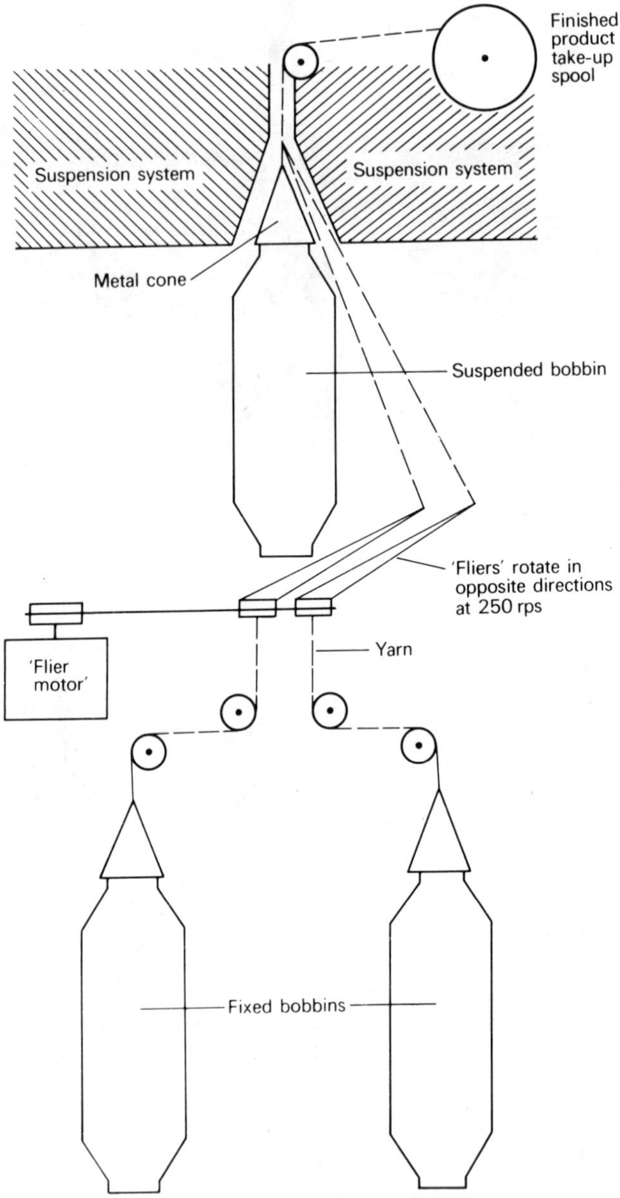

Fig. 5.41(a) Magnetically suspended yarn package for the formation of a plaited sewing cotton: schematic of the apparatus.

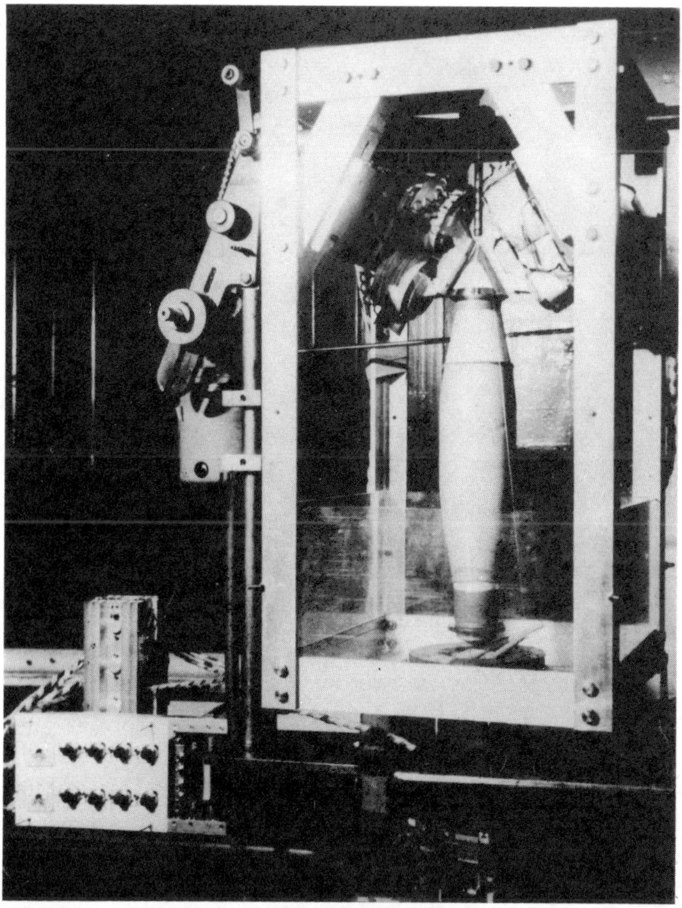

Fig. 5.41(b) Magnetically suspended yarn package for the formation of a plaited sewing cotton: prototype of the suspended yarn package.

laboratories, using the *I*-magnets. The controlled DC electromagnetic system is possibly the only one capable of operating at both high and low speeds. The eddy current drag was at one time considered to be another major drawback unless laminated rail could be used. This seems to have been overcome by the use of a large number of small magnets, thus retaining the rail in a magnetised state. The problem of power collection at speed is being tackled in the Emsland project by using a set of superconducting magnets on board the vehicle and an air cored synchronous motor on the track. It is believed that a small battery system on the vehicle will be sufficient to provide suspension and refrigeration power. The track winding will consist of cable wound in cement concrete slots, as used on the experimental track at Erlangen. It is safe to say that the problems of this technology in relation to transportation systems are not those of the technology itself but those of suitable propulsion units. Linear induction motors leave a great deal to be

Fig. 5.42 A magnetically suspended 1.6 m diameter steel drum weigh 8 tonne for centrifuge application. The drum is driven by an arch-shaped linear induction motor at a maximum speed of 375 rpm.

desired in terms of their weight together with any power conditioning equipment and power collection, whereas trackwound linear motors which would almost certainly be linear synchronous motors could make the track construction prohibitively expensive. In this connection it is worth mentioning that the cost of track and infrastructure is lower in hilly and mountainous regions where conventional railways would require tunnelling. The cost of any new advanced ground transportation scheme is likely to be determined primarily by the track costs and this in turn will be the principal factor determining whether such systems will replace aircraft in Europe by the turn of the century. Due to the exceptionally heavy investment required, unless all the European countries join forces and work towards an unified high speed ground transportation system for Europe, in addition to the railway network, the prospects will remain rather slim.

Bibliography

1 Geary, P. J., *Magnetic and electric suspensions.* SIRA Research Report No. R314, Brit. Sci. Instr. Res. Assc. (1964)

2 Earnshaw, S., On the nature of the molecular forces which regulate the constitution of the luminferous ether. *Trans. Camb. Phil. Soc.,* 7, pp. 97–112 (1842)

3 Braunbeck, W., Free suspension of bodies in electric and magnetic fields. *Zeitschrift für Physik,* **112**, 11, pp. 753–63 (1939)

4 Evershed, S., A frictionless motor. *Journal IEE,* 29, pp. 743–94 (1900)

5 Faus, H. T., *Magnetic suspension*. US Patent No. 2315 408, General Electric (US) (1943)

6 de Ferranti, S. Z., *Magnetic suspension for electricity meter spindles*. British Patent No. 590 292, Ferranti Ltd (1947)

7 *Magnetic bearings for aerospace applications*. General Electric (US) Advanced Technology Labs, Schenectady, NY, Tech. Doc. Report No. ASD-TDR-63-474 (1963)

8 Polgreen, G. R., *System of transportation.* British Patent No. 867 045 (1959)

9 Polgreen, G. R., Transport possibilities with magnetic suspension. *Electrical Times* (26 August 1965)

10 Polgreen, G. R., *New Applications of Modern Magnetics*. Macdonald Press, London (1966)

11 Polgreen, G. R., Guided land transport. *Proc. Inst. Mech. Eng.,* **181**, 3G, pp. 145–50 (1966–7)

12 Polgreen, G. R., Railway with magnetic suspension. *Engineer,* 226, pp. 632–6 (1968)

13 Polgreen, G. R., The ideal magnet–fully controllable permanent magnet for power and transport. *Electronics and Power*, 127, p. 233 (1971)

14 Magnetic vehicle suspension foreseen. *Westinghouse Engineer,* 25, pp. 95–6 (1965)

15 Forgacs, R. L., Evacuated tube versus jet aircraft for high speed transportation. *Proc. IEEE*, **61**, 5, pp. 604–16 (1973)

16 Baran, W., Optimising a permanent magnet suspension system for high speed ground transport. *Zeitschrift für angewandt Physik*, **32**, 3, pp. 216–18 (1971)

17 McCaig, M., Permanent magnets for repulsion systems. *Electrical Review*, 169, p. 425 (1961)

18 Bahmanyar, H., and Ellison, A. J., Application of permanent magnets to the suspension of surface guided vehicles. *Proc. IEE Conf. on Linear Electric Machines,* Publication No. 120, pp. 203–9 (1974)

19 Voigt, H., Electrodepamische Schwingungsdampfung im Permanentmagnet-Schwebesystemen. *Wiss Ber AEG-Telefunken,* 47, pp. 15–20 (1947)
20 Becker, J. J., Permanent magnets. *Scientific American*, **223**, 10, pp. 92–100 (Dec. 1970)
21 Braunbeck, W., Free suspension of diamagnetic bodies in magnetic fields. *Zeitschrift für Physik*, **112**, 11, pp. 764–9 (1939)
22 Braunbeck, W., Electrically or magnetically freely suspended bodies. *Umschau,* **53**, 3, pp. 68–70 (1953)
23 Boerdijk, A. H., Technical aspects of levitation. *Phillips Research Report*, **11**, 1, pp. 45–56 (1956)
24 Boerdijk, A. H., Levitation by static magnetic fields. *Phillips Technical Review*, **18**, 4, pp. 125–7 (1956)
25 Meissner, W., and Ochsenfeld, R., *Naturwissenschaften*, **21**, 78 (1933)
26 Arkadiev, V., Hovering of a magnet over a superconductor. *Journal of Physics*, Moscow, **9**, 2, p. 148 (1945)
27 Simon, I., Forces acting on superconductors in magnetic fields. *Journal of Applied Physics*, **24**, 1, pp. 19–24 (1953)
28 Culver, W. H., and Davis, M. H., *An application of superconductivity to inertial navigation.* Rand Corp., Santa Monica, Calif., Report No. R363, ASTMA ref. 249 458 (Jan. 1957)
29 Powell, J. R., *The magnetic road.* IEEE Paper No. 63-RR4 (April 1963)
30 Powell, J. R., and Danby, G. R., *High speed transport by magnetically suspended trains.* ASME Publication No. 66 WA/RR5 (Dec. 1966)
31 Guderjahn, C. A., Wipf, S. L., Fink, H. J., Boom, R. W., MacKenzie, E., Williams, D., and Downey, T., Magnetic suspension and guidance for high speed rockets by superconducting magnets. *Journal of Applied Physics*, **40**, 5, pp. 2133–40 (1969)
32 Borcherts, R. H., Davies, L. O., Reitz, J. R., and Wilkie, D. F., Baseline specifications for a magnetically suspended high speed vehicle. *Proc. IEEE,* 61, pp. 569–78 (1973)
33 Thornton, R. D., Design principles for magnetic levitation. *Proc. IEEE*, 61, pp. 586–98 (1973)
34 Ellison, A. J., and Bahmanyar, H., Surface guided transport systems of the future. *Proc. IEE,* **121**, 11R, pp. 1224–48 (1974)
35 Ohno, E., Iwamato, M., and Yamada, T., Characteristics of superconducting magnet suspension and propulsion for high speed trains. *Proc. IEEE,* 61, pp. 579–86 (1973)
36 Coffey, H. T., Barber, T. W., and Chilton, F., Magnetic suspension and guidance of high speed vehicles. *Journal of Applied Physics,* **40**, 5, p. 2161 (1969)
37 Ooi, B. T., and Banaker, M. H., Passive and active damper windings for repulsive magnetic levitation system. *Trans. IEEE on Magnetics*, **13**, 5, pp. 1418–20 (Sept. 1975)
38 Green, A. H., *et al,* Linear synchronous motor control of maglev vehicle ride quality. *Trans. IEEE on Magnetics,* **10**, 3, 431–4 (Sept. 1974)
39 Outsuka, T., and Kyotani, Y., Superconducting levitation high speed ground transportation project in Japan. *Trans. IEEE on Magnetics,* **11**, 2, pp. 608–12 (March 1975)

40 Yamamura, S., Magnetic levitation technology of tracked vehicles; present status and prospects. *Trans. IEEE on Magnetics,* **12**, 6, pp. 874–8 (Nov. 1976)

41 Tang, C. H., Harold, W. J., and Chu, R. S., A review of Magneplane project. *Trans. IEEE on Magnetics,* **11**, 2, pp. 625–6 (March 1975)

42 Reitz, J. R., and Borcherts, R. H., US Dept of Transportation program in magnetic suspension (repulsion concept). *Trans. IEEE on Magnetics,* **11**, 2, pp. 615–18 (March 1975)

43 Eastham, A. R., and Atherton, D. L., Superconducting maglev and linear synchronous motor developments in Canada. *Trans. IEEE on Magnetics,* **11**, 2, pp. 627–32 (March 1975)

44 Slemon, G. R., The Canadian maglev project on high speed interurban transportation. *Trans. IEEE on Magnetics,* **11**, 5, pp. 1478–83 (Sept. 1975)

45 Eastham, A. R., and Rhodes, R. G., The levitation of high speed trains using superconducting magnets. *Second International Symposium on Electromagnetic Suspension,* Southampton University, England (July 1971)

46 Rhodes, R. G., Mulhall, B. E., and Abel, E., Maglev vehicle oscillations and damping mechanisms. *Proc. IEE Conf. on Guided Land Transport,* Publication No. 117, pp. 214–20 (Aug. 1974)

47 Rhodes, R. G., *et al,* A superconducting maglev test facility for high speed transport. *Proc. 6th International Cryogenic Engineering Conf.,* Grenoble, IPC Sci. Tech. Press (May 1976)

48 Gutberlet, H. G., The German magnetic transportation program. *Trans. IEEE on Magnetics,* **10**, 3, pp. 417–20 (Sept. 1974)

49 Uranker, L., Survey of basic magnetic levitation research in Erlangen. *Trans. IEEE on Magnetics,* **10**, 3, pp. 421–4 (Sept. 1974)

50 Anschütz-Kaempfe, H., *Gyroscopes.* British Patent No. 193 397 (1923)

51 Orkress, E. C., Wroughton, D. M., Comentez, G., Brace, P. H., and Kelly, J. C. R., Electromagnetic levitation of solids and molten metals. *Journal of Physics,* **23**, 12, p. 1413 (1952)

52 Bedford, B. D., Peer, L. H. B., and Tonks, L., The electromagnetic levitator. *General Electric Review,* 42, p. 246 (1939)

53 Laithwaite, E. R., Electromagnetic levitation. *Proc. IEE,* **112**, 12, pp. 2361–75 (1965)

54 Laithwaite, E. R., Eastham, J. F., Bolton, H., and Fellows, T. G., Linear motors with transverse flux. *Proc. IEE,* **118**, 12, pp. 1761–7 (1971)

55 Eastham, J. F., and Laithwaite, E. R., Linear motor topology. *Proc. IEE,* **120**, 3, pp. 337–43 (1973)

56 Eastham, J. F., and Laithwaite, E. R., Linear induction motors as magnetic rivers. *Proc. IEE,* **121**, 10, pp. 1099–108 (1974)

57 Laithwaite, E. R., Magnetic or electromagnetic; the great divide. *Electronics and Power,* **19**, 4, pp. 310–12 (August 1973)

58 Eastham, J. F., *The magnetic river.* Paper presented to the town meeting of the SRC (Advanced Ground Transport Panel), London (24 Feb. 1978)

59 Pfann, W. G., and Hagelbarger, D. W., Electromagnetic suspension of a molten zone. *Journal of Applied Physics,* **27**, 1, pp. 12–18 (1956)

60 Skinner, D. J., and Edwards, D. R., Discussion at a meeting on *Superconducting Cables,* IEE, Savoy Place, London (25 Jan. 1978)

61 Nordsieck, A. T., Principles of the electric vacuum gyroscope. *American Rocket Society Guidance Control and Navigation Conf.*, Stanford, Calif., Paper No. 1963–61 (Aug. 1961)

62 Knobel, H. W., Electric vacuum gyro. *Control Engineering*, **11**, 2, pp. 70–3 (Feb. 1964)

63 Gillinson, P. J., Denhard, W. G., and Frazier, R. H., *A magnetic support for floated inertial instruments.* MIT Instrument Lab. Report No. R277 (April 1960)

64 Frazier, R. H., Gillinson, P. J., and Oberbeck, G. A., *Magnetic and Electric Suspensions.* MIT Press (1974)

65 *Investigation of magnetic and electric forces for rotating shaft suspension.* Tech. Doc. Report No. ASD-TDR-62-441, Aeronautical System Div. Airforce Systems Command, Wright Patterson Airforce Base, Ohio (May 1962)

66 Cambridge Thermionic Corp., Magnetic bearing minimises friction. *Electronic Design*, **11**, 7, pp. 32–3 (March 1963)

67 Cambridge Thermionic Corp., *Magnetic Suspension Apparatus.* US Patent No. 1401 514 (1975)

68 Hagihara, S., The performance and stability of a magnetic suspension device using a tuned LCR circuit. *Proc. IEE*, **125**, 2, pp. 153–6 (1978)

69 Kaplan, B. Z., Dynamic stabilisation of tuned circuit levitators. *Trans. IEEE on Magnetics*, **12**, 5, pp. 556–69 (1976)

70 Kaplan, B. Z., Estimation of mechanical transients in tuned circuit levitators by employing steady state impedances. *Journal of Applied Physics*, 47, 1, pp. 78–84 (1976)

71 Jayawant, B. V., and Rea, D. P., New electromagnetic suspension and its stabilisation. *Proc. IEE*, **115**, 4, pp. 549–54 (1968)

72 Kaplan, B. Z., Recent developments in magnetic levitation. *Electronics Letters*, **6**, 8, p. 70 (April 1970)

73 Kaplan, B. Z., Analysis of a method for magnetic levitation. *Proc. IEE*, **114**, 11, pp. 1801–4 (1967)

74 Graeminger, B., *Electromagnetic Suspension Devices.* British Patent Nos 24 499 (1912) and 24 541 (1912)

75 Anschütz-Kaempfe, H., *Gyroscopic Apparatus.* US Patent No. 1589 039 (1923)

76 Kemper, H., *Overhead suspension railway with wheelless vehicles employing magnetic suspension from iron rails* (Schwebebahn mit raderlosen Fahrzeugen die an eisernen Fahrschienen mittels magnetischer Felder schwebend entlang geführt werden). German Patent Nos 643 316 (1937) and 644 302 (1937)

77 Kemper, H., Suspension by electromagnetic forces: a possibility for a radically new method of transportation (Schwebende aufhangung durch electromagnetische Kraft; eine Möglichkeit für eine grundsatlich neue Fortbewegungsart). *Electrotechnische Zeitschrift*, **59**, 15, 00. 391–5 (1938)

78 Holmes, F. T., Axial magnetic suspension. *Rev. Sci. Instr.*, **8**, 11, pp. 444–7 (1937)

79 Beams, J. W., High rotation speeds. *Journal of Applied Physics*, **8**, 12, pp. 795–806 (1937)

80 Tournier, M., and Laurenceau, P., Magnetic suspension of windtunnel model. *Recherche Aeronautique*, 59, pp. 21–6 (1957)

81 ONERA, Magnetic suspension of windtunnel models. *Engineer,* **210**, 5463, pp. 607–8 (1960)

82 Clemens, P. L., and Cortner, A. H., *The suspension of windtunnel models; bibliography.* Arnold Engineering Center Tech. Doc. Report No. AEDC-TDR-63-20 (Feb. 1963)

83 Covert, E. E., and Finston, M., Magnetic balance and suspension systems for use with windtunnel models. *Progress in Aerospace Science,* 14, Chapter 2, pp. 27–107, Pergamon Press (1973)

84 Judd, M., and Goodyear, M., *Magnetic suspension system.* Annual Report of the Dept of Aeronautics, Southampton University, England (1965)

85 Wilson, A., and Luff, B., *The development design and construction of a magnetic suspension system for the RAE 7″ x 7″ hypersonic wind tunnel.* Royal Aircraft Establishment, Farnborough, Tech. Report No. TN66248 (1966)

86 Chrisinger, J. E., Tilton, E. L., Parkin, W. J., Coffin, J. B., and Covert, E. E., Magnetic suspension and balance system for windtunnel application. *Journal of the Royal Aeronautical Soc.,* 67, pp. 717–24 (1963)

87 Crain, C. D., *Design and initial calibration of a magnetic suspension system for windtunnel models.* Arnold Engineering Development Center Tech. Doc. Report No. AEDC-TDR-65-187 (1965)

88 Jenkins, A. W., and Parker, H. M., Electromagnetic support arrangement with 3-dimensional control. *Journal of Applied Physics,* supplement to Vol. 30, p. 2385 (April 1969)

89 Dukes, T. A., and Zapata, R. W., Magnetic suspension with minimum coupling effects for windtunnel models. *Trans. IEEE* (AES-1), 1, pp. 20–8 (1965)

90 Silver, K. R., and Henderson, J. G., One component magnetic support and balance for windtunnel models. *Journal of Aircraft,* **6**, 5, p. 398 (1969)

91 Kilgore, R. A., and Hamlet, I. L., Some aspects of an aircore single coil magnetic suspension system. *Summary of ARL Symposium on Magnetic Windtunnel Model Suspension and Balance Systems,* edited by F. L. Daum, No. 66-0133 (1966)

92 Jayawant, B. V., Sinha, P. K., Wheeler, A. R., Whorlow, R., and Willsher, J., The development of a 1 ton magnetically suspended vehicle using controlled DC electromagnets. *Proc. IEE,* **123**, 9, pp. 941–8 (1976)

93 Gottzein, E., and Cramer, W., Critical evaluation of multivariable control techniques based on maglev vehicle design. *Proc. IFAC Symposium on Multivariable Technological Systems,* New Brunswick (1977)

94 Nakamura, S., Development of high speed surface transport system (HSST) of Japan Air Lines. *Trans. IEEE on Magnetics,* **15**, 6, pp. 1428–33 (Nov. 1979)

95 Nakamura, S., Takeuchi, Y., and Takahashi, M., Experimental results of ferromagnetic levitation system of Japan Air Lines HSST-01/02 vehicles. *Trans. IEEE on Magnetics,* **15**, 6, pp. 1434–6 (Nov. 1979)

96 Bevir, M. K., *The stability of electromagnetic levitation systems for solid bodies.* UKAEA Culham Lab. preprint No. CLM-P458 (May 1976)

97 Rutherford Lab. and Culham Lab., Preliminary report and brochure for the Royal Society Conversazione (1976)

98 Homer, G. J., Rendle, T. C., Walters, C. R., Wilson, M. N., and Bevir, M. K., A new method for stable levitation of an iron body using superconductors. *Journal of Physics* D, 10, pp. 879–86 (1977)

99 Ross, J. A., Romag transportation system. *Proc. IEEE,* **61**, 5, pp. 617–20 (1973)
100 Edwards, J. D., and Antably, A. el, Segmented rotor linear reluctance motors with large airgaps. *Proc. IEE,* **125**, 3, pp. 209–14 (1978)
101 Eastham, J. F., Iron cored linear synchronous machines. *Electronics and Power,* **23**, 3, pp. 239–42 (1977)
102 Backers, F. T., A magnetic journal bearing. *Phillips Technical Review,* **22**, 7, pp. 232–8 (1961)
103 McCaig, M., Permanent magnets for repulsion devices. *Swift and Levick Bulletin* (1967)
104 Bahmanyar, H., *Novel transport system with particular reference to contact free suspensions.* Ph D Thesis, London University (1973)
105 London, F., *Superfluids.* Dover (1961)
106 Kammerlingh Onnes, H., Disappearance of electrical resistance of mercury at helium temperatures. *Konink Akad. Wetensch. Amsterdam Proc.,* **14**, Communication No. 122b, pp. 113–15 (June 1911)
107 Mathias, B. T., Superconductivity. *Scientific American,* **197**, 5, pp. 92–103 (1957)
108 Kunzler, J. E., Superconductivity in high magnetic fields at high current densities. *Reviews in Modern Physics,* **33**, 4, pp. 501–9 (Oct. 1961)
109 Arkadiev, V., A floating magnet. *Nature,* **160** 4062, p. 330 (1947)
110 Scott, R. B., *Cryogenic Engineering,* pp. 338–42. Van Nostrand (1959)
111 Mendelssohn, K., *Cryophysics,* Chapter 6. Interscience (1960)
112 Shoenberg, D., *Superconductivity,* pp. 16–20. CUP (1960)
113 Harding, J. T., and Tuffias, R. H., *The cryogenic gyro.* California Inst. of Technology Jet Propulsion Lab. Technical Release No. 34-100, ASTMA ref. AD249458 (Aug. 1960)
114 Buchold, T. A., *Gyroscope.* US Patent No. 3044 309 (1962)
115 Cryogenic gyro debuts. *Chemical and Engineering News,* **38**, 5, p. 34 (1 Feb. 1960).
116 New cryogenic research puts superconductivity to work. *Iron Age,* **185**, 4, pp. 102–4 (Jan. 1960)
117 Cryogenic devices on the way. *Electronics,* **33**, 6, pp. 32–3 (Feb. 1960)
118 Cryogenic gyro. *Mechanical Engineering,* **82**, 3, pp. 78–9 (March 1960)
119 Project spin works toward development of cryogenic gyroscope. *Electrical Engineering,* **79**, 4, p. 344 (April 1960)
120 Reitz, J. R., Forces on moving magnets due to eddy currents. *Journal of Applied Physics,* **41**, 5, pp. 2067–71 (1970)
121 Reitz, J. R., and Davis, L. C., Force on a rectangular coil moving over a conducting slab. *Journal of Applied Physics,* **43**, 4, pp. 1547–53 (1972)
122 Davis, L. C., Drag force on a magnet moving near a thin conductor. *Journal of Applied Physics,* **43**, 10, pp. 4256–7 (1962)
123 Reitz, J. R., Borcherts, R. H., Davis, L. C., Hunt, T. K., and Wilkie, D. F., *Technical feasibility of magnetic levitation as a suspension system for HSGT vehicles.* Ford Motor Co. Technical Report No. FRA-RT-72-40 produced under contract DOT-FR-10026, US Dept of Transportation (Feb. 1972)
124 Borcherts, R. C., and Davis, L. C., Force on a coil moving over a conducting surface including edge and channel effects. *Journal of Applied Physics,* **43**, pp. 2418–27 (1972)

125 Elliott, D. G., *Matrix analysis of linear induction machines.* US Dept of Transportation Report No. FRA-ORD-75-77 (Sept. 1975).

126 Tinkham, M., AC losses in superconducting magnetic suspension for high speed transportation. *Journal of Applied Physics,* **44,** 5, pp. 2385–90 (1973)

127 Laithwaite, E. R., *Induction Machines for Special Purposes.* Newnes (1966)

128 Jayawant, B. V., *Induction Machines.* McGraw-Hill (1968)

129 Yamamura, S., *Theory of Linear Induction Motors.* Wiley (1972)

130 Ooi, B. T., and White, D. C., Traction and normal forces in linear induction motors. *Trans. IEE Power App. and Syst.,* PAS-89, pp. 638–45 (1970)

131 Parsch, C. P., and Gibson, J. P., The aircored linear synchronous motor. Basic operation and test facilities at Erlangen. *Colloquium on Advanced Group Transportation Schemes,* IEE Digest No. 1979/27, pp. 3.1–3.4 (1979)

132 Kolm, H. H., and Thornton, R. D., The magneplane: guided electromagnetic flight. *Proc. IEEE Conf. on App. Superconductivity* (1972)

133 Kolm, H. H., and Thornton, R. B., Electromagnetic flight. *Scientific American,* **229,** 4, pp. 17–25 (Oct. 1973)

134 Slemon, G. R., Burke, P. E., and Turton, R. A., A linear synchronous motor for high speed ground transport. *Digest of INTERMAG Conf.,* pp. 19–6/1 (May 1974)

135 Eastham, A. R., (editor), Superconducting magnetic levitation and linear synchronous motor propulsion for high speed ground transportation. *Canadian Inst. of Guided Transport Report* No. 75–5 (1975)

136 Laithwaite, E. R., *High speed ground transport.* Royal Institution discourse 1973 (2 Feb. 1973)

137 Barwell, F. T., and Laithwaite, E. R., Application of linear induction motors to high speed ground transport. *Proc. Inst. Mech. Eng.,* 181, 3G (1966–7)

138 Chirgwin, E. M., US linear induction motor research vehicle program. *Proc. IEE Conf. on Linear Electric Machines,* Publication No. 120, pp. 236–43 (1974)

139 West, J. C., and Hesmondhalgh, D. E., Analysis of thick cylinder induction machines. *Proc. IEE,* 109C, pp. 172–81 (1962)

140 West, J. C., and Jayawant, B. V., Self oscillating linear induction motor. *Proc. IEE,* 109A, pp. 292–300 (1962)

141 Bamji, P., Some applications of linear electric machines for handling molten aluminium. *Proc. IEE Conf. on Linear Electric Machines,* Publication No. 120, pp. 68–76 (1974)

142 Cashmore, D. H., *Electromagnetic levitation.* M Sc Thesis, Manchester University (1963)

143 Hesmondhalgh, D. E., Unpublished work, Manchester University (1955)

144 Lovell, W. V., Electromagnet removes non-ferrous materials. *Electronics,* **28,** 9, p. 146 (Sept. 1955)

145 Polonis, D. H., Butters, R. G., and Parr, J. G., Some techniques for melting reactive metals. *Research,* 7, p. 272 (1954)

146 Scheibe, W., Uber einige Probleme aus die Hock metallurgie. *Metall,* 7, p. 151 (1953)

147 Weisberg, L. R., Levitation of molten gallium, indium, gold and tin. *Rev. Sci. Instr.,* 30, p. 315 (1959)

148 Brisley, W., and Thornton, B. S., Electromagnetic levitation calculation for axially symmetric systems. *British Journal of Applied Physics,* **14,** 10, pp. 682–6 (Oct. 1963)

149 Freeman, E. M., and Laithwaite, E. R., Unbalanced magnetic push. *Proc. IEE,* **115**, 4, p. 538 (1968)
150 Eastham, J. F., and Laithwaite, E. R., Linear motor topology. *Proc. IEE,* **120**, 3, pp. 337–43 (1973)
151 Rajanathan, C. B., Lowther, D. A., and Freeman, E. M., A study of the Xi-core transverse flux plate levitator. *Proc. IEE,* **127B**, 3, pp. 183–9 (1980)
152 Laithwaite, E. R., (editor), Transport without wheels. *Elek. Science,* Chapter 11, p. 295 (1977)
153 Freeman, E. M., and Lowther, D. A., Normal forces in single sided linear induction motors. *Proc. IEE,* **120**, 12, pp. 1499–506 (1973)
154 Williams, F. C., Laithwaite, E. R., and Piggott, L. S., Brushless variable speed induction motors. *Proc. IEE,* **104A**, 6, pp. 102–18 (1957)
155 Eastham, J. F., and Balchin, M. J., Pole change linear induction motors. *Proc. Conf. IEE,* Publication No. 120, pp. 9–14 (1974)
156 Aylwin, D. G., *Design of controlled DC magnetic suspension system.* M Phil Dissertation, University of Sussex (1977)
157 Katz, R. M., Nene, V. D., Ravera, R. J., and Skalski, C. A., Performance of magnetic suspensions for high speed vehicles operating over flexible guideways. *Trans. ASME,* **96**, G2, pp. 204–12 (1974)
158 Popp, K., and Schiehlen, W., Dynamics of magnetically levitated vehicles on flexible guideways. *Proc. IUTAM Symposium on the Dynamics of Vehicles on Roads and Railway Tracks,* Delft (1975)
159 Meisinger, R., Control systems for maglev vehicles riding over flexible guideways. *ibid.*
160 Gottzein, E., Cramer, W., Ossenberg, F. W., and Roche, C., Optimal control of a maglev vehicle. *ibid.*
161 Gottzein, E., and Lange, B., Magnetic suspension control systems for MBB high speed train. *Automatica,* **11**, pp. 271–84 (1975)
162 Gottzein, E., Brock, K. H., and Pferfferl, J., Control aspects of a magnetic levitation high speed test vehicle. *Automatica,* **13**, pp. 201–23 (1977)
163 Meisinger, R., *Beiträge zur reglung einer magnet Schwebebahn auf elastichem Fahrweg.* Dr Ing Dissertation, Tech. Univ. München (1977)
164 Jayawant, B. V., Sinha, P. K., and Aylwin, D. G., Feedback control systems for DC electromagnets in passenger carrying vehicles. *Int. Journal of Control,* **24**, 5, pp. 627–39 (1976)
165 Jayawant, B. V., and Sinha, P. K., Low speed vehicle dynamics and ride quality using controlled DC electromagnets. *Automatica,* **13**, 6, pp. 605–10 (1977)
166 Sinha, P. K., Analytical and design aspects of passenger carrying vehicles using controlled DC electromagnetic suspension. *Proc. 4th MVTS Symposium,* New Brunswick, pp. 573–81 (1977)
167 Gottzein, E., Meisinger, R., and Miller, L., Anwendung des 'Magnetischen Rades' in hochgeschwindigkeits magnet Schwebebahnen. *ZEV-Glasers Annalan,* **103**, 5, pp. 227–32 (1979)
168 Hazlerigg, A. D. G., Design of a multivariable controller for a magnetically supported vehicle. *Proc. IEE Conf. on Control Aspects of Guided Land Transport,* Publication No. 117, pp. 233–9 (1974)

169 Wheeler, A. R., *Aspects of design and construction of DC magnetic suspension systems,* Chapter 4, p. 60, D Phil Dissertation, University of Sussex (1975)

170 Lion, K. S., Non-linear twin T-network for capacitive transducers. *Rev. Sci. Instr.,* **33**, 5, pp. 535–6 (1962)

171 Hodkinson, R. L., *The suspension and stabilisation of an electric train by electromagnets.* M Sc Dissertation, University of Sussex (1972)

172 Whorlow, R. J., *Design of magnetic suspension systems and control of vehicles in networks.* D Phil Dissertation, University of Sussex (1978)

173 Edwards, J. D., Internal report, University of Sussex

174 Whorlow, R. J., Internal report, University of Sussex

175 Yamamura, S., and Ito, T., Analysis of speed characteristics of attractive magnets for magnetic levitation of vehicles. *Proc. INTERMAG. Conf.,* Paper 28 (1975)

176 Willsher, J., *The development of passenger carrying vehicles using controlled DC electromagnets,* Chapter 2, pp. 48–57. M Phil Dissertation, University of Sussex (1978)

177 Jayawant, B. V., and Wheeler, A. R., *Electromagnetic Suspension Railway Assembly.* British Patent No. 1557 864 (1979)

178 Crede, C. E., *Vibration and Shock Isolation.* John Wiley, New York (1951)

Index